OUR SOLAR SYSTEM

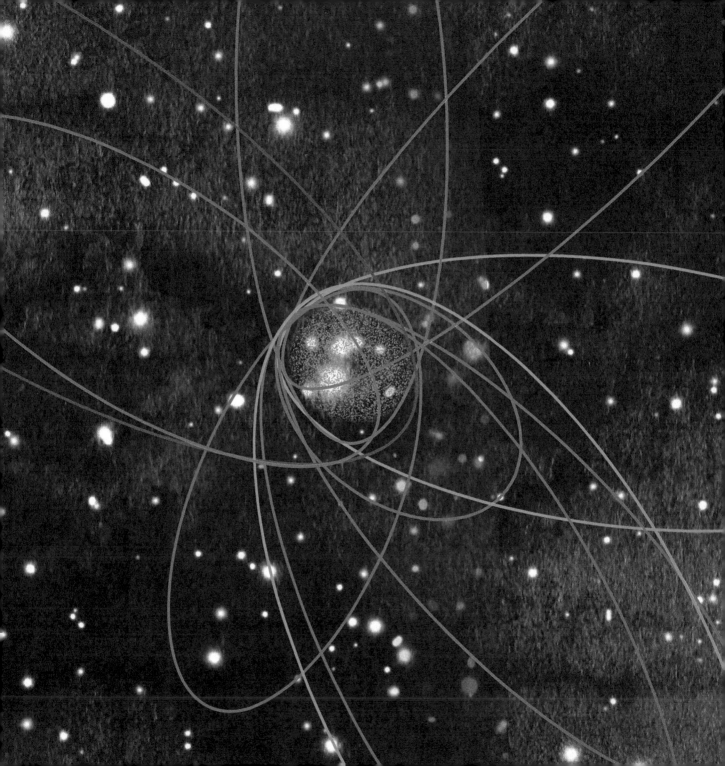

OUR SOLAR SYSTEM

AN EXPLORATION OF PLANETS, MOONS, ASTEROIDS, AND OTHER MYSTERIES OF SPACE

LISA REICHLEY

ROCKRIDGE PRESS

Interior and Cover Designer: Eric Pratt
Art Producer: Tom Hood
Editor: Aarti Kelapure

Illustrations © Hans Bennewitz, 2020, cover and pp 16, 17, 25, 27, 28, 48, 58, 75, 76, 128, 133, 148, 158. Photographs Mohd Nadly Aizat Nudri/Alamy, p. viii; Jacek Bakutis/Alamy, p. 8; Alan Dyer/VWPics/Alamy, p. 9; SPL/Science Source, p. 10; NASA pp 14, 24, 35, 38, 43, 54, 62, 65, 69, 72, 79, 84, 87, 96, 100, 105, 109, 111, 113, 122, 136,; J Marshall - Tribaleye Images/Alamy, p. 31; MARKA/Alamy, p. 49; Dinodia Photos/Alamy, p. 50; QAInternational/Science Source, p. 90; NASA Image Collection/Alamy, pp 93, 139; NG Images/Alamy, p. 103; NASA/Jet Propulsion Laboratory, p. 116; Diego Barucco/Alamy, pp 142, 145; Dmitry Larichev/Alamy, p. 150; Author photo courtesy Tammy Snyder.

ISBN: Print 978-1-64739-913-9 | eBook 978-1-64739-914-6
R0

This book is dedicated to my mom, Rita,
who always believed I could achieve anything
I set my mind to. Thanks, Mom!

CONTENTS

The Milky Way viewed from Earth.

CHAPTER 1

OUR SOLAR SYSTEM

Does the world seem like a really big place to you? Well, it is, until you think about how Earth fits into the bigger picture. It's just one planet in our **solar system**, which is just a little dot in our galaxy, one of more than 100 billion galaxies in the universe. That's right—100 billion galaxies! Now does the world seem like a really big place? Probably not quite as much. Let's take a closer look at where we fit in in the universe.

OUR SOLAR SYSTEM

BY THE NUMBERS

PLANETS: 8

DWARF PLANETS: AT LEAST 5

PLANETARY MOONS: MORE THAN 200

ASTEROIDS: MORE THAN 1,000,000

COMETS: MORE THAN 4,000

This book is about our solar system. Have you ever wondered why we call it that? The word *solar* refers to our Sun, which is a star. It's the only star in our solar system, which also contains the planets, dwarf planets, **satellites**, and other objects in their orbits. It's a system because all the bodies found around the Sun move and work together. These bodies can function as a system because they're caught up in the exceptional gravitational force of the Sun.

Our solar system is one of many such systems found in the Milky Way galaxy. A **galaxy** is a collection of millions or billions of stars, gas, dust, and other bodies all held together by gravity. Experts believe there are more than 100 billion galaxies visible from Earth. As technology improves and we learn more about the solar system, this number will likely grow. The Milky Way is visible without a telescope. On a clear, moonless night, in an area with little light pollution, you can see the Milky Way in a swath of stars across the sky. It's easiest to see in July and August.

Scientists estimate that there are about 250 billion stars in the Milky Way. What does 250 billion look like? That number may seem hard to imagine, so think of it this way: one square inch of grass contains about 50 individual blades. If each blade of grass (including the Sun) is a star, the Milky Way is as big as about 600 football fields! The Milky Way is too large to photograph, but artists have created depictions of our spiral-shaped galaxy. Our entire solar system is just a dot on the image, and the individual planets are just too small to show.

The size of the solar system is measured in **astronomical units**, or AU. The AU is based on the distance from the Sun to Earth, which is 93 million miles. So when we say Neptune is 30 AU from the Sun, this means it's 30 times farther from the Sun than Earth is, or 2,790 million miles away. The diameter of the solar system—that little dot in the Milky Way—is 100,000 AU, or 930 billion miles.

Our solar system is in the Orion Arm of the Milky Way, about halfway between the center and the outer edge of the galaxy. We call our solar system *the* solar system because it's where we live. This is similar to how we refer to our moon as *the* Moon. Earth's moon is the only unnamed moon in the solar system. We call it the Moon because it's ours.

We know there are other planets and moons in our solar system, but are there other planets around the other stars in our galaxy? Yes, there are. In fact, astronomers have been searching for other Earthlike planets around other stars in our galaxy. They've already identified about 2,500 such systems in the Milky Way, but they believe there could be millions. We won't be traveling to them anytime soon, however. They're simply too far away.

We're going to explore our solar system in this book, but not just the Sun, the planets, and their moons. We'll also look closely at what else exists within this system, and we'll seek to answer some important questions: Where did the solar system come from? Did the Sun always exist, and more important, will it always exist? Why is Pluto no longer a planet? What are comets made of? How do planets and other bodies get their names?

So, what are we waiting for? Let's start exploring!

WHERE DID OUR SOLAR SYSTEM COME FROM?

Scientists believe that our solar system formed approximately 4.5 billion years ago from a dense cloud of **interstellar** gas and dust: gas and dust that forms between (*inter*) the stars (*stellar*). What kinds of gas and dust were there? Most of the gas present was hydrogen. Hydrogen is the simplest element on the periodic table and one of the essential building blocks of life. The dust present in this interstellar cloud would have been microscopic. Microscopic dust is only visible when you have a large amount in a small space.

There are multiple theories about how the solar system was formed. The theory that most scientists believe today is the Solar Nebular Theory.

THE SOLAR NEBULAR THEORY

To understand this theory, you must first understand that gravity is present with all matter and the gravitational force of an object is directly dependent upon its mass. This is why, for example, if you were to visit the Moon, you would only weigh one-sixth of what you weigh on Earth. The Moon is less massive, so its gravitational force is much less.

The Solar Nebular Theory states that our solar system began as a slowly rotating **nebula**, or interstellar cloud of gas and dust. As the nebula rotated, the gravity present caused the collapsing of the cloud into a flattened disc shape. Gases and dust that were close to one another slowly clumped together. With every increase in mass came an increase in gravitational force, which, in turn, attracted more particles.

According to this theory, the Sun, the planets, their moons, and other solar system bodies formed at approximately the same time. Hydrogen gas collected to form the Sun. As the Sun grew, its mass caused it to collapse in on itself. When this happened, the hydrogen atoms began the process of **nuclear fusion**. Nuclear fusion happens when atoms fuse, or merge, together. In this case, hydrogen atoms fused together to form helium. When helium is formed, a great deal of energy is released in the form of heat and light. This increase in temperature at the center of the solar system caused many of the other gases in the solar system to burn off.

Our solar system wasn't born overnight. In fact, scientists believe its formation process occurred over a period of 100 million years. No one was able to witness the formation of our solar system, but scientists have based these hypotheses on their observations of other nebulae.

WHAT EXISTS BETWEEN THE BODIES IN OUR SOLAR SYSTEM?

When we discuss the solar system's bodies, such as the planets and moons, we often note things like size, temperature, and composition. But this discussion rarely focuses on what happens after leaving a planet's atmosphere. Is there something between the atmospheres of two neighboring planets or is it a complete void of space? There are lots of things, actually! Between bodies in our solar system there is cosmic dust, solar wind, magnetic fields, and electromagnetic radiation.

COSMIC DUST

If you ever worried about your bedroom being dusty, don't feel bad. The solar system is full of dust. Most of the cosmic dust is limited to our orbital plane, the flat area that connects the object being orbited (the Sun) with the objects orbiting it (everything else in the solar system). Dust extends from the Sun out to the farthest reaches of the solar system.

SOLAR WIND

The Sun emits a stream of **solar wind** in all directions. Nuclear fusion from inside the sun and magnetic activity near its surface result in charged particles, and those particles make up solar wind. Solar wind carries a lot of energy that could be very harmful to us. Luckily, we have a shield of protection—the Earth's magnetic field. While we can't see solar wind, we can see it interacting with the magnetic field.

MAGNETIC FIELDS

Earth has a **magnetic field** that extends beyond our atmosphere and into space. Magnetic fields aren't composed of matter. Rather, a magnetic field is the area over which a magnet exerts a force. Within the Earth lies a liquid outer core, composed primarily of iron and nickel. As the Earth spins, this liquid moves around, generating an electric current, which in turn generates a magnetic field.

The Sun, Earth, and many other planets have magnetic fields of varying strength. A body's magnetic field is affected by the composition of the core and the rate at which the body spins.

ELECTROMAGNETIC RADIATION

All matter creates, reflects, and absorbs different **electromagnetic wavelengths**. Light is one kind of electromagnetic wave. Electromagnetic waves are made of electric and magnetic fields, not matter. Unlike sound waves, electromagnetic waves do not need matter to travel through. Depending on the wavelength, they can be very harmful or completely harmless.

Another name for electromagnetic waves is *radiation*. Radiation refers to the way in which the energy is transferred. Energy radiates, or travels outward, from a central location: in this case, the Sun. We can see a portion of the electromagnetic waves given off by the Sun in the form of light. We can feel another wavelength, infrared, as warmth from the Sun. And you've probably heard of a harmful radiation called ultraviolet rays, which can cause sunburn or even skin cancer. These are all examples of electromagnetic waves that travel throughout space.

WHAT CAN WE SEE FROM EARTH?

What do you see when you look up at the sky? During the day, the Sun, clouds, and sometimes even the Moon are visible. But what about at night? That depends on a variety of factors: time of year, your geographical location, the amount of light pollution where you are, and the weather. The Moon, stars, perhaps a planet or two will be visible on most nights. If you're lucky, you may have seen the Milky Way galaxy across a late summer sky. Perhaps you've even seen a **meteor shower** or an **eclipse**. You may have even seen evidence of solar wind without realizing it.

THE AURORA BOREALIS

An aurora visible from Earth.

You've learned that the Sun gives off solar wind, which is a constant stream of charged particles. Solar wind is invisible unless it's in the Earth's magnetic field. As mentioned earlier, this field is generated by the liquid outer core of the Earth. Like a bar magnet, the Earth's magnetic field is strongest at its poles. When the solar wind reaches the Earth's magnetic field, it's pulled by the magnetic force to either the North Pole or South Pole. As the solar wind's charged particles move toward the poles and into the Earth's atmosphere, they interact with the gases in our atmosphere. This interaction results in the beautiful light displays known as the Aurora Borealis in the Northern Hemisphere and the Aurora Australis in the Southern Hemisphere. So, while we can't see solar wind or Earth's magnetic field, we *can* see how they interact with each other, and the result is one spectacular show!

ZODIACAL LIGHT AND THE *GEGENSCHEIN*

Is the cosmic dust ever visible from Earth? Yes, it is. There are two different night sky phenomena associated with cosmic dust: **zodiacal light** and the ***gegenschein***. Before we discuss what these are, let's first look at the Earth's location relative to the Sun, zodiac constellations, and cosmic dust.

The Sun takes the same basic path across the sky each day. This path is called the **ecliptic**. The zodiac constellations also take this same ecliptic path. The zodiac constellations are found on the same plane as the Sun and planets.

Zodiacal light visible from Earth.

The gegenschein visible from Earth.

The word *gegenschein* is a German word that means "counterglow." The *gegenschein* is similar to the zodiacal light, though it's seen much less frequently. While the zodiacal light illuminates cosmic dust around the Sun, the *gegenschein* illuminates cosmic dust on the opposite side of the Earth as the Sun. As a result, it's a very faint glow that can only be seen in the absence of moonlight or light pollution.

Earth's orbit, the Sun, cosmic dust, and the constellations of the zodiac are all on this same orbital plane. And all these things, minus Earth, of course, take the ecliptic path across our sky.

Zodiacal light is also referred to as "false dusk" or "false dawn." Just after the Sun sets, or before it rises, you can sometimes see a cone- or pyramid-shaped area of light coming up from the horizon along the ecliptic. This is the Sun's rays reflecting off the cosmic dust surrounding the Sun, on the same plane as the Sun and the zodiac constellations. This light was named the zodiacal light because it's on the ecliptic, so you might think it's light from the zodiac constellations.

HOW DID WE LEARN ABOUT OUR SOLAR SYSTEM?

Because the Sun, the Moon, the stars, Mercury, Venus, Mars, Jupiter, and Saturn are all visible with the naked eye, people have known about them for centuries. Even before the invention of the telescope, people were able to track the movement of some stars and planets across the night sky.

HISTORICALLY SPEAKING

The first telescope patent was awarded to Dutch eyeglass-maker Hans Lippershey in 1608. The following year, the famed Italian astronomer Galileo Galilei made his own telescope, similar in design to others being produced at that time. Galileo is credited with being the first to use his telescope to observe and note features of the Moon. He later would also discover Jupiter's four largest moons, which are now collectively referred to as the *Galilean satellites*. Although Saturn had already been identified before the invention of the telescope, Galileo was the first to observe it through a telescope and notice its rings.

It was initially believed that the Earth was the center of our solar system and that the Sun, the Moon, and the planets (Mercury, Venus, Mars, and Jupiter) revolved around it. This theory is called the **geocentric** model (*geo:* Earth; *centric:* at the center). It was developed by Greek mathematician Ptolemy in the 2nd century, but it wasn't widely accepted until the 16th century.

The **heliocentric** model—the idea that the Sun is at the center of the solar system—was first proposed by Polish astronomer Nicolaus Copernicus in 1543. Although we now know this theory to be true, it wasn't immediately accepted at the time. In fact, almost a century later in 1632, after studying the movement of Venus, Galileo agreed with Copernicus that the planets did, in fact, orbit the Sun. The following year Galileo was arrested and convicted for having beliefs that went against the Catholic Church. According to the Church's interpretation of the Bible, the Earth was the center of the solar system. Galileo spent the rest of his life under house arrest, and it would be hundreds of years before the Church admitted that the Sun was, in fact, the center of the solar system.

Uranus was discovered in 1781 by English astronomer William Herschel. It was the first planet to be discovered by telescope. Neptune was then discovered in 1846 using mathematics. Later in the book, you'll learn that several people were given credit for its discovery, as they all arrived at the same conclusion working independently from one another.

MORE RECENTLY

Technology developed over the last century has improved the telescope and has allowed us to see more of the solar system. Today there are many observatories where scientists and often the public can look through enormous, powerful telescopes to observe the solar system. Space telescopes such as the Hubble, which launched in 1990, orbit the Earth. They allow astronomers to view space unimpeded by clouds or light pollution. In 2006, the New Horizons spacecraft was launched with the mission of capturing images of Pluto, its moons, and other distant objects. It's the first spacecraft to photograph these distant worlds. An even more powerful telescope, the Webb telescope, is expected to launch in 2021. Named after former NASA administrator, James Webb, this telescope is 100 times more powerful than the Hubble. It will orbit the Sun about a million miles away from Earth.

We're constantly learning about the solar system, and much of it is still largely unexplored.

CHAPTER 2

THE SUN

THE SUN

MASS: 4.385×10^{30} POUNDS
(99.86 PERCENT OF THE MASS OF THE SOLAR SYSTEM)

DIAMETER: 865,370 MILES (109 TIMES THE DIAMETER OF EARTH)

AVERAGE TEMPERATURE: 9,941°F

DISTANCE FROM EARTH: 1 AU (93 MILLION MILES)

AGE: 4.6 BILLION YEARS

If the Earth were the size of a Ping-Pong ball, the Sun would be the height of two and a half refrigerators. It's *that* massive.

The Sun is at the center of our solar system. Composed primarily of hydrogen and helium, it makes up 99.86 percent of the mass of our solar system, which is why we call it the *solar* system and not, say, the *Jupiter* system. Remember, the word *solar* refers to the Sun. Our entire system of planets, moons, and other bodies is controlled by the Sun. The Sun's mass holds all the planets in orbit. If the Sun's gravity didn't exist, each planet would fly off in a straight line into space. The Sun's gravity is the force that holds the planets in their orbits.

OUR SUN: A STAR

The Sun is a star—the only star in our solar system. It's a G-type main sequence star, sometimes called a **yellow dwarf**. Most stars are main sequence stars. Main sequence stars are composed of hydrogen that's undergoing nuclear fusion, creating helium. While our sun is technically medium-sized, it is actually larger than most stars in the Milky Way. At the same time, there are some stars that are enormous compared with our sun.

Our entire solar system revolves around this *one* star. There are probably billions of other such systems in the giant, spinning spiral that is the Milky Way galaxy. Remember: Our entire solar system is a mere dot among 250 billion stars swirling around in the Milky Way galaxy.

WHAT'S THE SUN MADE OF?

The Sun is mostly made up of hydrogen and helium. We generally think of hydrogen and helium as gases, but because the Sun is so hot, its hydrogen and helium are **plasma**. Plasma is the fourth state of matter (along with solid, liquid, and gas) that forms when excessive amounts of heat cause the electrons to escape the atom and the result is charged particles, also called **ions**. Together hydrogen and helium make up 98 percent of the Sun's composition. Carbon, nitrogen, and oxygen make up another 1.5 percent, and the remaining 0.5 percent includes neon, iron, silicon, magnesium, and sulfur.

The layers of the sun.

THE SUN'S INTERIOR

The mass of the Sun not only holds all the planets in orbit, it also provides the force behind its massive production of energy. The Sun's enormous mass puts an immense amount of pressure on its center, or core. The pressure at the core generates temperatures as high as 27 million degrees Fahrenheit, making it the hottest spot in our solar system. The heat and pressure in the core are responsible for the fusion of hydrogen into helium. This can only occur in the core because it's the only layer where this amount of pressure, and as a result this heat, exist.

The core is one of three main layers that make up the inside of the Sun. As you move outward through the layers of the Sun, several things happen. There's less pressure, the density decreases, and the temperature decreases.

That heat generated within the core travels outward. It first passes through the **radiative zone**. The radiative zone is sandwiched between the core and the **convective zone**. Because the radiative zone is farther from the center of the Sun, there's less pressure and less heat. In fact, there's not enough pressure for fusion to occur here. The energy produced by the fusion in the Sun's core radiates—or spreads out—through this layer, which is how it got its name.

Energy then travels through the Sun's convective zone, which is named for the convection currents that exist there. Convection is a type of heat transfer in which heat is carried by a moving fluid. The fluid moves in a swirling or churning motion because some areas are hotter than others. We have the same type of heat transfer within the mantle of the Earth, where hot magma rises and cool magma sinks. Rather than traveling directly outward, the heat, energy, and plasma of the Sun get caught up in convection cells, which are like churning areas of movement. It might help you understand convection cells if you think about the heaters in homes, which are often near the floor. When the heat comes on in a cool room, it causes movement of the air in the room. These are convection cells. The same thing happens in the convective layer of the Sun. There is churning because it's hot on the inside and cooler on the outside. This movement likely generates the Sun's complex magnetic field.

THE SUN'S EXTERIOR

The atmosphere of the Sun is broken down into three layers: the photosphere, the chromosphere, and the corona.

Energy is released from the core, travels through the radiative zone and the convective zone, and finally arrives at the **photosphere** (*photo:* light), where it's visible in the form of light. At 10,000 degrees Fahrenheit, this is the coolest layer of the Sun.

The chromosphere and corona are only visible during a solar eclipse. The **chromosphere** has a faint red color (*chromo:* color). The **corona** (*corona:* crown) is the outermost layer of the Sun's atmosphere. The temperature rises when passing from the photosphere to the chromosphere to the corona, even though these layers are farther away from the core. The Sun's temperature goes from 27 million degrees Fahrenheit at the core, down to 10,000 degrees Fahrenheit at the surface, and then back up to 2 million degrees Fahrenheit in the corona.

SUNSPOTS

The photosphere is where we see cooler areas on the Sun's surface called **sunspots**. Because these areas are cooler, they appear darker than the area around them, and the result looks like a spot on the Sun. Sunspots are small regions where the Sun's magnetic field is unusually strong. Sunspots occur in pairs—a northern and a southern sunspot. When gases flow along these field lines, we see a **prominence**, burning gases that come out from the Sun's surface, often in a loop. The magnetic field poles rising to the Sun's surface slow down the movement in the convective zone below it. Because the convective zone is slowed, the heat on the surface is not steadily replaced, making it cooler.

Sunspots occur in cycles that last approximately 11 years. The Sun always has sunspots, but it goes through periods when the number of them changes. Watching the movement of sunspots has taught us about the rotation of the Sun. Scientists can watch sunspots move on the Sun and keep track of how long it takes them to do a full rotation. By watching the spots, scientists discovered that the Sun rotates faster at the equator and slower near its poles. The equator rotates once every 24 days, while it takes more than 30 days for a single rotation at the poles.

SOLAR FLARES

The Sun's rotation can cause the magnetic field lines at the Sun's surface to become twisted and tangled. When these lines get too mixed up, they'll rearrange. When this happens, there can be a sudden release of energy in the form of X-rays, a flash of light, radio waves—a whole mess of electromagnetic radiation. This is called a **solar flare** and it's visible as a flash of light.

CORONAL MASS EJECTIONS

Coronal mass ejections (**CMEs**) are often associated with solar flares, but they're two different things. While a solar flare is a release of energy, a CME is a release of actual *stuff*: particles (protons and electrons). These are the same particles that make up solar wind. A CME is like a great big cough of the Sun's solar wind. Because CMEs are actual particles with mass, they move more slowly than solar flares. The energy released by a solar flare will reach Earth in a matter of minutes, while the particles of a CME will take days.

HOW DO SOLAR FLARES AND CMES AFFECT EARTH?

Both solar flares and CMEs can affect us here on Earth. Because a solar flare is a burst of electromagnetic waves, a strong one can disrupt radio communication. When the particles from a CME reach our magnetosphere, they are pulled toward the poles and interact with our atmosphere to create beautiful lights called *auroras*. Solar wind does the same thing. Both solar wind and CMEs are caused by particles being emitted from the Sun. Scientists believe that if a CME were large enough, it could affect our power grids and cause widespread power outages.

THE SUN FROM EARTH

Have you ever wondered why you need to wear sunscreen on a sunny day? It's because certain rays from the Sun, called ultraviolet rays, can be very harmful to your skin. The ultraviolet wavelength is just one wavelength on the electromagnetic spectrum. It immediately follows the violet wavelength. Because the ultraviolet wavelength is just a little shorter than the violet wavelength, it transfers a bit more energy. It turns out that this little bit of extra energy is very damaging to our skin cells. Sunscreen protects us from the energy of the ultraviolet rays.

We cannot see the ultraviolet wavelength, but many animals can, including birds and bees. Humans are only able to see the visible portion of the electromagnetic spectrum: red, orange, yellow, green, blue, indigo, and violet. (Your science teacher may have taught you the acronym ROY G. BIV to help you remember the colors of the spectrum in order.) All these colors together make up white light. The visible portion of the electromagnetic spectrum is what allows us to see. If you're looking at something orange, it appears orange because it reflects the orange wavelength. Plants are green because they reflect the green wavelength. Black absorbs all wavelengths of visible light. If you wear a black shirt outside on a hot day, you will become very hot because the shirt is absorbing all the energy being transferred by the visible wavelengths. If you were to wear a white shirt, it would reflect all the colors, rather than absorb them, and you would be much cooler.

WHAT COLOR IS THE SUN?

When you were little, you probably colored the Sun in a coloring book. What crayon did you use? Orange? Yellow? You probably never thought to use a white crayon to color the Sun, but maybe you should have. Although the Sun appears to be yellow or orange in the sky, it's actually white. You might say, "Wait, we just learned that the sun is a *yellow* dwarf star!" It's true that the Sun is called a yellow dwarf, but that name is a bit misleading. The Sun appears yellow or orange to us because we're seeing it through the filter of our atmosphere.

If you've ever worn sunglasses, you know that they change the way things appear. Well, the atmosphere does the same thing. The Sun's light passes through the atmosphere the same way light passes through sunglasses, and this interaction affects the way color is perceived. The light from the Sun is scattered by the molecules that make up our atmosphere. This effect is known as **Rayleigh scattering**. Most of our atmosphere is nitrogen and oxygen, and these molecules are fairly close in size to each other. They both scatter wavelengths around the blue wavelength of the visible spectrum. That's why the sky is blue. When you look at the Sun in the sky (which you should never do directly), the wavelengths of all the blue and colors near blue, such as green and indigo, are scattered away. The wavelengths left behind are red, orange, and yellow. These colors give the Sun its yellow appearance.

You might be tempted to point out that the sky isn't blue at sunrise or sunset, and you'd be right. During sunrise or sunset, the Sun's rays are passing through the atmosphere at a different angle, and this changes the filter through which they pass. The bottom layer of the atmosphere is the most dense. By cutting through this layer at a steeper angle, more yellow and orange wavelengths are filtered out, leaving the Sun with a more reddish appearance than other times of the day.

SOLAR RADIATION

The wavelength just longer than the red wavelength is called infrared. This wavelength has a little bit less energy than the red one. We can't see this wavelength, but we can feel it. When you feel the warmth of the Sun on your face, you're feeling the infrared waves. Unlike the rays from the ultraviolet wavelength, infrared waves are not harmful.

All the electromagnetic waves given off by the Sun, including the infrared and ultraviolet waves, make up **solar radiation**. We can feel warmth from the infrared waves, we can see some colors on the visible portion of the electromagnetic spectrum, and we can be harmed by damaging ultraviolet light.

When solar radiation reaches our atmosphere, several things can happen. Some energy is immediately reflected off the atmosphere. Some energy is absorbed by the atmospheric gases. And still more is reflected back to space by Earth's surface or the cloud tops. The rest is absorbed into bodies of water, the atmosphere, or the Earth itself. Solar radiation is important to life on Earth. For example, solar panels can absorb solar radiation and convert it to usable energy. Some solar radiation is absorbed and used in the process of photosynthesis, in which plants convert sunlight into energy so that they can grow.

SOLAR ECLIPSES

An eclipse is when the light from a body in our solar system is obscured or blocked by another body. On Earth, we can observe both solar and **lunar** eclipses. Lunar eclipses, as we'll discuss later, can be viewed by anyone on a night with clear weather.

A solar eclipse occurs when the Moon moves between the Earth and the Sun, blocking the Sun's light. These are far less common than lunar eclipses. That's because the Moon's shadow cast on Earth is very small. Why? Because the Moon is small compared with the Earth. If a mouse were standing between you and a light, its shadow would be very small on you. Almost the entire half of you facing the light would not be in the mouse's shadow. For the same reason, a solar eclipse is only visible to people within a small area of the Earth.

A solar eclipse takes about three hours from start to finish, but the time of **totality**, when the Moon is blocking the entire Sun, is only about 7½ minutes. During the totality, you can see the Sun's corona.

HOW TO VIEW A SOLAR ECLIPSE

Never ever look directly at the Sun, not even during an eclipse. This is dangerous and can cause serious damage to your eyes. If you want to see a solar eclipse in real time, build a pinhole projector.

You will need:

- **tape**
- **a long empty box, such as a cereal box**
- **a pin or a needle**
- **scissors**

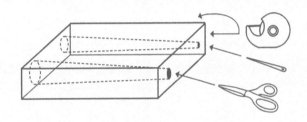

Find a long empty box, and tape it shut. On the left-hand side of the box's top, about an inch from the edge, make a small hole using a pin or a needle. This hole is for the sunlight to enter. On the right-hand side of the cereal box top, make a larger hole with scissors. This hole is for you to look through, so it should be about the same size as your eye.

During the eclipse, take your projector outside. Make sure you're facing away from the Sun, and hold your eye up to the larger hole. The sunlight will enter through the pinhole and you can safely view the eclipse.

THE LIFE OF OUR SUN

Stars have a life cycle. They're born, they live a long life, and, eventually, they die. Don't worry, though—this won't happen to our Sun anytime soon.

Our Sun was born 4.6 billion years ago. The Sun is a main sequence star, which means that it spends most of its life undergoing nuclear fusion. About 90 percent of a star's life is spent in this stable phase. Our Sun began as a sphere of hydrogen gas. Once enough pressure gathered in the center, fusion began. This fusion continues today, but eventually, the Sun will stop fusing hydrogen in its core.

At the end of the main sequence stage, a star is no longer stable, and it begins to die. The core is no longer fusing hydrogen into helium. At this point, the core is made up of *burnt* helium, which will contract and burn the leftover hydrogen surrounding it. This causes the star to expand to huge proportions! At this point, the star is a red giant. Our Sun will become a red giant in about 5 billion years. That's when the Sun will swallow Mercury, Venus, and possibly Earth. Once the outward force of energy escaping the Sun has been exhausted, it will collapse. At this stage of its life, it is called a white dwarf.

The length of a star's life depends on its size. You might think a larger star would live longer because it has more fuel, but that's not true. Actually, more massive stars also have more massive pressure, so they burn through their fuel more quickly and die sooner than smaller stars.

CHAPTER 3

THE INNER SOLAR SYSTEM

So far, we've looked at the general characteristics of the solar system. We've also learned about the Sun. Now, it's time to move outward from the Sun to start to take a closer look at what else makes up our solar system.

Let's begin with the inner solar system. This region of the solar system is made up of the inner planets: Mercury, Venus, Earth, and Mars; their moons; and the asteroid belt. We can see all the inner planets in the night sky. People have known about this group of planets for thousands of years.

The inner planets are often collectively referred to as **terrestrial planets** because they have characteristics that are similar to Earth's. They're not similar enough for us to live on any of them, but these planets do have characteristics in common with ours, including a solid surface and a rocky, metal composition. The most common metal found in the terrestrial planets is iron.

The inner planets share other characteristics, as well. They're all very small and much closer together than the outer planets. Earth is the largest inner planet. Venus, whose diameter is 95 percent of Earth's, is a close second. The diameter of Mars is about half that of Earth's. Mercury, the smallest inner planet, has a diameter that's about 38 percent of Earth's.

While the inner planets vary in size, they are very small compared with the outer planets. Earth is the largest inner planet, and Neptune is the smallest outer planet. But you could still fit 57 Earths into Neptune!

The inner planets have few or no moons—Mercury and Venus have no moons, Earth has one, and Mars has two moons that are more like asteroids—and inner planets don't have any rings.

The inner planets tend to rotate more slowly than the outer planets, but they travel around the Sun faster. This gives the inner planets longer days but shorter years. While it takes Neptune 165 Earth years to orbit the Sun, Mercury accomplishes the task in 88 Earth days. And it's not just because Mercury has a smaller orbit: Mercury is also moving at 10 times the speed of Neptune.

MERCURY
BY THE NUMBERS

DIAMETER: 3,032 MILES
DISTANCE FROM SUN: 0.39 AU
KNOWN MOONS: NONE
LENGTH OF DAY: 58 EARTH DAYS
LENGTH OF YEAR: 88 EARTH DAYS
TEMPERATURE RANGE: −290°F TO 800°F

The inner solar system also includes the asteroid belt. The asteroid belt is a disc found between the orbits of Mars and Jupiter that contains millions of small rocky bodies, called asteroids, that orbit the Sun. Asteroids are left over from the creation of the solar system.

MERCURY

Named after the Roman messenger god, Mercury is the closest planet to the Sun. It's also the swiftest planet in the solar system, completing an orbit around the Sun in just 88 Earth days. Mercury has a thin atmosphere, which gives it the greatest temperature differential—the difference between the highest and the lowest temperatures on a single planet—in the solar system. And because of its orbital and rotational speeds, Mercury is the only place in the solar system with a double sunrise!

FEATURES

Mercury fits in well with the inner planets. It has a solid surface, no moons, and no rings. It's made up primarily of iron and other heavy metals. It has a cratered surface that's gray in color. Because Mercury has no wind, rain, or volcanoes, these craters don't erode or get covered up. Mercury is about one-third the size of Earth, just slightly larger than our Moon.

UNDERSTANDING MERCURY'S ORBIT

Mercury's orbit is the most eccentric of any planet. It's not really a circle, like the other planets' orbits. Imagine pinching a circle just a bit: that's what Mercury's orbit is like. It's also slightly tilted (approximately seven degrees) compared with other planets' orbits.

The point in Mercury's orbit when it comes closest to the Sun is called the **perihelion** (*peri:* near; *helion:* from *helios*, the Greek word for *sun*). The point in Mercury's orbit where it's farthest from the Sun is the **aphelion** (*ap:* away from).

Mercury is rotating very slowly but orbiting the Sun very quickly. Mercury completes only three rotations for every two orbits, which affects how long a day is on Mercury.

Look at the chart **Mercury: By the Numbers** (page 31). Here, you will see that a single Mercury day is 58 Earth days. This is a **sidereal day** on Mercury. The word *sidereal* means "in reference to the stars." From the perspective of the stars, it takes 58 Earth days for Mercury to rotate one time.

If you were *on* Mercury, a day would take 176 Earth days. Why the difference? Because if you were on Mercury, a day is measured by the position of the Sun. This is called a **solar day** because it is based on where the Sun appears in the sky. Mercury is rotating counterclockwise and traveling counterclockwise around the Sun, and it takes a long time for the Sun to get back to the same position. This means that on Mercury, one day is twice as long as a year!

DOUBLE SUNRISE

Mercury's rotational speed is always the same, but its orbital speed varies, depending on its position in orbit. When Mercury travels along the perihelion portion of its orbit, it is moving about 25 percent faster than during the rest of its orbit. This increase in speed means it is now moving *faster* in its orbit than it's rotating. If you were in a certain place on Mercury, the Sun would appear to move *backward* in the sky during this time! Don't worry though—this only occurs over a period of a few Earth days. After Mercury moves out of this portion of its orbit, the Sun quickly catches up to its original position. So, in some places on Mercury, the Sun would rise, travel backward, and then rise again. A double sunrise—only on Mercury!

SOLAR SIZE

Because Mercury is closest to the Sun, the Sun appears larger. On Earth, the Sun measures about 0.5 degrees in the sky. If you were to stand on Mercury, the average size of the Sun (because it fluctuates from perihelion to aphelion) would measure 1.4 degrees. That's almost three times as large as it appears on Earth! Because Mercury is so close to the Sun, and the Sun is so large, the Sun is six times brighter on Mercury than what we experience on Earth.

IT'S IMPORTANT TO WEAR YOUR JACKET

Think of a planet's atmosphere like a jacket: It holds heat in and protects the planet, just like your jacket holds heat in and protects you on a cool day. You might expect Mercury to be the hottest planet because it's closest to the Sun, but that's not the case. You see, Mercury isn't wearing a jacket. Its neighbor Venus is slightly farther away from the Sun, but it has on a *very* heavy jacket, making it the hottest planet.

Mercury's barely existent atmosphere is constantly blasted by solar wind. When Mercury is facing the Sun, temperatures can rise to a sweltering 800 degrees Fahrenheit. But because it is not wearing a jacket, only the surface of the planet can absorb the heat. Without a jacket, the temperature on the side of the planet facing away from the Sun will plummet to a frigid −290 degrees Fahrenheit. That's a difference of 1,090 degrees, the largest temperature differential in the solar system.

In addition to affecting the temperature, the atmosphere of a planet also protects its surface. For example, most objects that enter Earth's atmosphere burn up entirely before reaching Earth's surface. Mercury doesn't have this layer of protection, and the result is a heavily cratered surface.

VIEWING MERCURY

Two spacecraft have gone to Mercury: once in 1974 and then again in 2008. Photographs taken during both journeys have revealed similarities between Mercury's surface and that of Earth's moon. Both are heavily cratered, but these photos show that Mercury has even more craters than our Moon and any other planet.

Although it is difficult, you can see Mercury from home with an amateur telescope. It's always found by looking in the same direction as the Sun. Because of its closeness to the Sun, the two are never separated by more than 27 degrees. The best time to view Mercury is just before sunrise or just after sunset, depending on its position in orbit. Mercury is easier to see when the Sun is below the horizon. If the Sun is above the horizon, its glare makes Mercury more difficult to see.

VENUS

BY THE NUMBERS

DIAMETER: 7,521 MILES
DISTANCE FROM SUN: 0.72 AU
KNOWN MOONS: NONE
LENGTH OF DAY: 243 EARTH DAYS
LENGTH OF YEAR: 225 EARTH DAYS
AVERAGE TEMPERATURE: 864°F

VENUS

Venus is named after the Roman goddess of love and beauty. But if you traveled to Venus, the experience would be neither lovely nor beautiful. In fact, you'd be flattened like a pancake!

FEATURES

Like Mercury, Venus has no moons and no rings. Its internal structure is very similar to Earth's. There's a core of iron surrounded by a mantle and covered with a rocky crust. While the rocks on the surface of Venus appear to be gray, if you were on Venus everything would have a yellow-orange tinge to it due to the filtering of the sunlight through the thick, Venusian atmosphere.

Venus has mountains, plains, and volcanoes. The highest mountain on Venus is Maxwell Montes, standing seven miles above the average height of the crust. More than 90 percent of Venus's surface is covered in lava flows and shield volcanoes (broad, gently sloping volcanoes that resemble a warrior's shield placed on the ground). The shield volcanoes on Venus are even broader and wider than those on Earth.

Most everything that enters the Venusian atmosphere burns up before reaching the surface. This is due to the high friction caused by the thick atmosphere. As a result of its atmosphere, and eruptions of its many volcanoes, Venus has fewer craters than Mercury.

THAT ATMOSPHERE!

Venus has the highest atmospheric pressure of any planet in the solar system. At Venus's surface, the atmosphere has 92 times the pressure of Earth's atmosphere. Just the pressure alone would make it impossible for humans to live there.

Venus's atmosphere is 96.5 percent carbon dioxide. By comparison, Earth's atmosphere is 0.04 percent carbon dioxide. Carbon dioxide is a greenhouse gas, which means that it readily absorbs the Sun's energy. So Venus's atmosphere absorbs energy from the Sun at a much higher rate than Earth's atmosphere.

In the last section we discussed how Mercury lacking an atmosphere was like the planet not wearing a jacket. While Mercury has no jacket, Venus has on the heaviest jacket in the entire solar system. Because of its thick atmosphere and the fact that it's the second planet from the Sun, Venus's atmosphere absorbs and traps the Sun's heat. The average temperature on Venus is a sweltering 864 degrees Fahrenheit, making it the hottest planet in the solar system.

Venus is also the brightest planet in the night sky. Clouds made of sulfuric acid reflect a large amount of the sunlight that strikes the planet. This, combined with our proximity to Venus, make it bright and easy to see, even without a telescope.

A DIFFERENT SPIN ON THINGS

Of all the planets, Venus's orbit comes closest to a perfect circle, but its rotation is a different story. Venus has a **retrograde** rotation, which means that it spins backward. Except for Uranus, which has its own weird rotation (as we'll discuss later), all other planets rotate counterclockwise. The retrograde rotation makes Venus different in several ways. For one, unlike on Earth, the Sun rises in the west and sets in the east. In addition, the retrograde spin has an effect on the length of a day. On Venus, one sidereal day is 243 Earth days. But because it is spinning in the opposite direction, one solar day on Venus would only be 117 Earth days. The Venusian solar day is shortened by the fact that it spins backward, unlike Mercury, where the solar day is extended.

VISITING VENUS

The United States and other countries have sent more than a dozen missions to Venus. Humans cannot land on Venus, but spacecraft have orbited the planet, and some have even landed to take measurements and photographs. For example, a robotic spacecraft measured the temperature and pressure of the atmosphere as it made its way to the surface. A separate spacecraft studied the rocks on the surface. These spacecraft didn't have long to transmit data back to Earth. Within an hour or so of landing, they melted and were crushed by the Venusian atmosphere!

EARTH'S SISTER

Humans could never land on Venus, much less live there. Why, then, is it referred to as Earth's *sister planet*? Despite Mars being a closer match to Earth, both in temperature and atmosphere, Venus is a closer match in size and distance from the Sun. These characteristics earned the planet the title of Earth's sister planet.

EARTH
BY THE NUMBERS

DIAMETER: 7,917.5 MILES

DISTANCE FROM SUN: 1 AU

KNOWN MOONS: 1

LENGTH OF DAY: 23 HOURS, 56 MINUTES, 4 SECONDS

LENGTH OF YEAR: 365.25 DAYS

AVERAGE TEMPERATURE: 58.3°F

EARTH

With its large oceans that regulate planetary temperatures, an ozone layer that absorbs ultraviolet rays, and a magnetic field that deflects solar wind, Earth is teeming with life.

EARTH'S ATMOSPHERE

Earth's atmosphere regulates temperature, distributes water around the planet, burns up meteors, absorbs ultraviolet rays, and allows for radio-wave communication. Nitrogen and oxygen make up 99 percent of the atmosphere. Earth's atmosphere can be broken down into five layers, each determined by the temperature changes that occur. Let's explore them.

THE TROPOSPHERE

The troposphere (*tropo:* change) is the bottom layer of the atmosphere. This is where all weather and 99 percent of water vapor are found. It is called the *sphere of change* because the weather is constantly changing. As you travel upward through the troposphere, it gets colder. The air also becomes thinner and less dense. This makes it colder and more difficult to breathe at the top of a mountain.

THE STRATOSPHERE

The stratosphere (*strato:* layer) contains the ozone layer, which is why the temperature increases here. The ozone layer absorbs the energy from the Sun's ultraviolet rays. This layer also contains the jet stream, a fast stream of moving air. Airplanes fly in the stratosphere because there's less friction due to fewer air molecules. If the jet stream is moving in the same direction as the airplane, it provides a nice tailwind, making the plane ride smoother and faster.

THE MESOSPHERE

The mesosphere (*meso:* middle) is the middle sphere of the Earth's atmosphere. This is the first layer where meteors, upon entering the atmosphere, will encounter enough friction to break apart. We call these broken meteors shooting stars. Have you ever seen one? If not, start looking. All over the planet, there are more than one million shooting stars every night!

THE THERMOSPHERE

The thermosphere (*thermo:* heat) is the warmest layer of the atmosphere. It is within the thermosphere that we see the Aurora Borealis or the Aurora Australis, depending on where you are. Radio waves bounce between this layer and the Earth's surface, making long-distance radio communication possible.

THE EXOSPHERE

The exosphere (*exo:* outside) is the outermost layer of the atmosphere. In this layer, there are fewer and fewer air molecules, and the atmosphere starts blending into space. The atmosphere does not end in one absolute spot. Think of a beach. There is no one spot where the sand ends completely. This is how the air molecules work in the exosphere. You might find a satellite orbiting the Earth in this layer.

SEASONS

Each year on January 4, the Earth is at perihelion, or the position in its orbit where it comes closest to the Sun. Aphelion is on July 4. We are about three million miles closer to the Sun at perihelion than we are at aphelion, but these distances have absolutely nothing to do with why we have seasons.

The Earth's axis, the imaginary line on which it rotates, is tilted on an angle of 23.5 degrees. The tilt points in the same direction throughout our entire orbit. The Earth is also a sphere. When the north pole is tilted in the direction of the Sun, the Sun's direct rays hit the top part of the sphere, north of the equator. Because the Sun's rays are concentrated when they strike at a 90-degree angle, the heat is concentrated, and the weather is warm. This is summer in the Northern Hemisphere. At the same time, the Sun's rays are striking south of the equator on a shallow angle, causing the Sun's heat to be spread over a larger area. The rays are spread so thin that there's minimal heat, making it cold. When the Sun's rays are at an angle like this, the Southern Hemisphere is experiencing winter.

LEAP YEAR

Look back at the **Earth: By the Numbers** chart (page 38) and you'll see that the length of the year is 365.25 days. But you've always thought that a year was 365 days, haven't you? In reality, after 365 days have passed, we have not yet completed an entire orbit. We're short by one-quarter of one day. After a second year, we're short by half a day. Then a third year goes by. Finally, after four years have passed, we are a full day short of where we began our orbit four years earlier. So we add a day to the calendar to make up for the change of position in orbit. That day is February 29, and it occurs every fourth year, called a leap year. If we didn't add this day to the calendar, eventually the seasons would shift to different months!

EARTH'S ROTATION

Just like you always assumed that a year is 365 days, you've also probably assumed that a day is 24 hours. Look again at **Earth: By the Numbers**. You'll see that a day is listed as 23 hours, 56 minutes, and 4 seconds. This is a sidereal day on Earth. From the viewpoint of the stars, this is how long it takes for the Earth to rotate exactly once. The problem is, once the Earth has rotated once, it has also moved in its orbit. The Earth needs to continue to rotate for an additional 3 minutes and 56 seconds for the Sun to be back in the same position. Here on Earth, we don't really care about the length of a sidereal day. We care about the length of a solar day. Why? Because we're *on* Earth. We want to know how long it takes the Sun to return to the same position.

OCEANS OF WATER

Seventy percent of the Earth's surface is water. Water has a very high specific heat, which means it has to gain a lot of energy to heat up, and it has to lose a lot of energy to cool down. This property helps regulate the temperatures on Earth. In other words, it keeps temperatures from fluctuating wildly. There's also a global conveyor belt, a current of ocean water that encompasses the Earth. Just as a conveyor belt at a factory transports goods as it moves, so, too, does the global conveyor belt transport water and energy, and this also helps distribute heat.

Long-range weather forecasts are often determined by oceanic temperatures. *El Niño* and *La Niña* are warmer-than-usual and cooler-than-usual temperatures in the Pacific. While the ocean temperature can't give the day-to-day forecast, the information helps us know the general weather conditions for the upcoming months.

THE MOON
BY THE NUMBERS

DIAMETER: 2,158.8 MILES

DISTANCE FROM EARTH: 238,855 MILES

SIDEREAL MONTH: 27.3 DAYS

SYNODIC MONTH: 29.5 DAYS

LENGTH OF ORBIT AROUND

EARTH: 27.3 DAYS

AVERAGE TEMPERATURE: −298°F

EARTH'S MOON

Most nights, we can look up and see the Moon shining in the night sky. But *moonlight* is actually just sunlight reflecting off the Moon's surface. Sometimes the Moon looks enormous, but in reality, it's only a quarter of the size of Earth. Because it has less mass than Earth, the Moon has a weaker gravitational pull. If you were on the Moon, you'd only weigh one-sixth of what you weigh on Earth!

THE MOON'S ORBIT

In **The Moon: By the Numbers** (page 43), you can see that the sidereal month is 27.3 days. According to the stars, it takes the Moon 27.3 days to rotate once. But wait! It takes 29.5 days for one complete lunar cycle (called a synodic month)—for the Moon to go from new moon to new moon. Why the difference?

It's true that, from the perspective of the stars, the Moon has rotated once in 27.3 days. However, the Earth has moved in its orbit. If Earth had stayed still, we would have completed a lunar cycle in these 27.3 days. But we did not stay still. Because Earth moved in its orbit around the Sun, from *our* perspective, the Moon hasn't returned to its beginning phase, and won't for two more days.

WHY DO WE ALWAYS SEE THE SAME SIDE OF THE MOON?

The Moon is in a **tidal lock** with Earth. What does this mean? Well, it means that because Earth's gravity has a stronger pull on the side of Moon closest to it, and a weaker pull on the far side of the Moon, the same side of the Moon is always facing the Earth. (For more about tidal locking, see the section on Charon, Pluto's largest moon, page 139.)

The Moon rotates exactly one time for each orbit it completes. To get an idea of what this looks like, put a chair in the middle of the room and have someone sit in it. Stand facing this person, just a couple feet from the chair. Hide something behind your back. Stepping sideways, go around the chair once, facing the chair the whole time. Make sure that the person in the chair can't see what you have behind your back.

Notice how, as you move around the chair, you face each of the walls in the room. This is because you are also rotating as you move. It just so happens that you're rotating at the same rate as you're moving around the chair. This is exactly how the Moon moves around Earth. The same side is always facing Earth. Not because the Moon is *not* rotating, but because it is rotating at the same rate it is moving around Earth.

The far side of the Moon is more heavily cratered than the side that faces us. One reason may be that once the Moon tidally locked with Earth, it protected the inside face of the Moon from being struck by meteoroids.

WHY ARE THE MOON AND THE SUN THE SAME SIZE IN THE SKY?

How large an object appears depends on two things: its actual size and how close you are to it. The Sun's diameter is about 400 times that of the Moon. The Sun is also 400 times farther from Earth than the Moon. Because these numbers are the same, both the Moon and the Sun appear to be the same size in the sky.

As discussed earlier, when the Moon is in front of the Sun during a solar eclipse, the entire Sun is blocked and we can observe the Sun's outer atmosphere, or corona. Even though the Moon is a fraction of the Sun's size, it still effectively blocks everything but the corona. However, the Moon is moving away from the Earth at a rate of about 1.5 inches a year. One day, this distance will affect its ability to eclipse the Sun. But don't worry: at a rate of 1.5 inches a year, it will be a *long* time before we stop having solar eclipses.

A MOON WITH MANY FACES

The Moon is a very popular celestial body, probably because we see it so clearly and so often. For centuries, people having been naming lunar events based on the Moon's color, position, size, and appearance, as well as the timing of its cycle. Here are a few of the more popular and well-known lunar events:

BLOOD MOON

A blood moon gets its name from a reddish tint that can sometimes be visible on the Moon. This only occurs during a lunar eclipse. Just before (and after) the Moon moves into Earth's shadow, the light hitting the Moon is filtered through Earth's atmosphere. Essentially, all the sunsets on Earth are projected onto the Moon's surface. With all the shorter wavelengths being filtered out by Earth's atmosphere, only red wavelengths remain, making the Moon look red.

BLUE MOON

The Moon never turns blue. The lunar cycle takes 29.5 days. A typical calendar month has either 30 or 31 days, except for February, of course. Because having two full moons in a single calendar month is rare, the second of these full moons is referred to as a blue moon. With only 28 days, February will never have one. Sorry, February!

SUPER MOON

The Moon's orbit is not a perfect circle. When the Moon is the closest to Earth as it will come, we say it is at **perigee**. When the Moon is as far as it travels from Earth, we say it is at **apogee**. The difference in these distances is about 26,000 miles. When the Moon is at perigee during a full moon, it appears larger than at other times in its orbit. We call this a super moon because it appears to be super-sized: about 14 percent larger than normal.

HARVEST MOON

The harvest moon is the full moon closest to the fall equinox. The fall equinox is a location in Earth's orbit when the tilt is not pointed toward or away from the Sun. Instead, it is perpendicular to the Sun's location. This marks the first day of autumn in the Northern Hemisphere and falls near September 21. The only astronomical significance is that the fall equinox marks the astronomical beginning of fall. The name *harvest* is used simply because this would be close to the time of the final harvest of the year.

VISITING THE MOON

The Moon is the only body in space to which humans have traveled. In 1969, Neil Armstrong became the first person to walk on the Moon. Armstrong was on the Apollo 11 mission, part of a series of NASA space missions with the goal of putting people on the surface of the Moon and returning them safely to Earth. In all, 12 people have walked on the Moon.

PHASES OF THE MOON

When the lunar cycle begins, we cannot see the lit side of the Moon. This is called a new moon because it marks a new cycle. Halfway through the lunar cycle, when we see the entire lit side of the Moon, we call it a full moon.

As the Moon travels around Earth, we see varying amounts of its lit side. During the first half of the lunar cycle, when we see more and more of the lit side each night, we say the Moon is **waxing**. During the second half of the lunar cycle, when we see less and less of the lit side each night, we say the Moon is **waning**. Waxing moons are visible in the evening, just after sunset. If you see the Moon out in the morning, it's passing through the waning phases.

Another characteristic we describe with the phase name is the shape of the Moon. If less than half the circle is illuminated, we say it's a **crescent** moon. If more than half the circle is lit, we say it's a **gibbous** moon.

The phases of the Moon.

LUNAR ECLIPSE

Why don't we have a lunar eclipse every time the Earth is between the Sun and the Moon? The three bodies must be aligned for us to see an eclipse, but the Moon's orbit is on a tilt, so this does not actually happen all that often. The Moon's orbit is tilted 5 degrees when compared with the Earth's orbit. The only times the Sun, Earth, and Moon are all on the same plane are close to the spring and fall equinoxes, March 21 and September 21, respectively.

A lunar eclipse is visible in the night sky wherever there's clear weather. Unlike a solar eclipse, which happens only in certain parts of the world, a lunar eclipse is visible for anyone on the side of the Earth that's experiencing night. This is because the Earth is much larger than the Moon.

The Moon is eclipsed when it moves into Earth's **umbra**, or *shadow*. Earth's umbra is large compared with the Moon. The area surrounding the umbra is called the **penumbra**, which is Latin for *almost shadow*.

A lunar eclipse as viewed from Earth.

MARS

BY THE NUMBERS

DIAMETER: 4,212.3 MILES
DISTANCE FROM SUN: 1.5 AU
KNOWN MOONS: 2
LENGTH OF DAY: 24 HOURS, 37 MINUTES, 22 SECONDS
LENGTH OF YEAR: 687 EARTH DAYS
AVERAGE TEMPERATURE: −80°F

MARS

Mars, also known as the Red Planet, is named after the Roman god of war. Filmmakers have set many science fiction movies on Mars because its surface is most like Earth's. Just like Earth, Mars completes one rotation in slightly more than 24 hours. Also, thanks to its 25-degree tilt, Mars has seasons, just like we do. With all these similarities, is Mars the next frontier for humans?

FEATURES

Mars is a terrestrial planet. It has a solid surface and is composed primarily of rock. Mars has two moons—Phobos and Deimos—but no rings. The diameter of Mars is slightly greater than half of Earth's diameter.

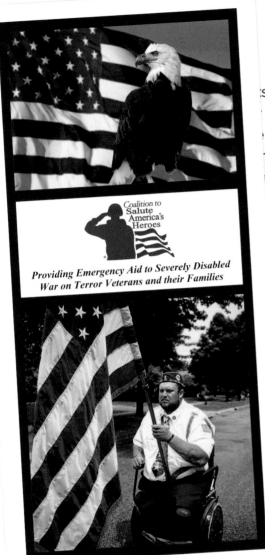

:s, is composed mainly of rock, with an iron core. on its surface, scientists believe that there was . When iron is exposed to water or oxygen, a to the oxygen, producing iron oxide, better Red Planet because the large amounts of iron appearance.

IOSPHERE

ercent of which is carbon dioxide. Although as and responsible for trapping heat on Venus, in for that to happen on Mars. In fact, Mars's it the density of Earth's atmosphere. To con- re being a planet's jacket, we might describe nd since Mars is the farthest inner planet from

s a different composition and density than ths. Remember, this filter affects how light and esult, on Mars, the sky looks red during the

ts surface, Mars has ice caps. The Mar- i's. On Earth, our ice caps are made of are made of frozen water and frozen

GLOBAL WINDSTORMS

During the day, the temperature on Mars can reach a balmy 70 degrees Fahrenheit at the equator. At the poles, it can dip down to −200 degrees Fahrenheit. As the surface of Mars becomes heated by the Sun in the daytime, the upper atmosphere remains cold. When the hot surface meets the cool atmosphere, the result is a windstorm.

Areas of warm air rise. Areas of cooler air sink. On Mars, these windstorms can last weeks, or even months, and can even be global events. Would they be dangerous to us if we were on Mars? Not really. Winds from these storms top out at about 60 miles per hour, which is pretty weak compared with the destructive force of hurricane winds here on Earth.

Because the surface of Mars is dry, windstorms create dust storms. These dust storms block out the sunlight and there is less temperature variation, so the source of energy is cut off. In essence, blowing dust up into the atmosphere works to calm the storm.

OLYMPUS MONS

Mars is home to Olympus Mons, the largest volcano in the entire solar system. Olympus Mons is a shield volcano. Because Mars doesn't have **tectonic plates**, its volcanoes are the result of **hot spots**. A hot spot occurs when hot magma melts through the crust, forming a volcano on the surface. On Earth, tectonic plates limit the size of a shield volcano because the moving plates shift the volcano away from the hot spot. On Mars, the volcanoes are never moved from their location above a hot spot, so they just continue to grow with each eruption. Because of this, the volcanoes on Mars are 10 to 100 times the size they are on Earth.

VISITING MARS

On May 30, 1971, NASA launched the Mariner 9 from Cape Canaveral, Florida. This spacecraft was the first ever to orbit another planet, Mars. Mariner 9 returned more than 7,000 photographs detailing the Martian surface, as well as its moons, Phobos and Deimos.

Viking 1 and Viking 2 missions launched from Cape Canaveral in 1975 and 1976, respectively. Each mission had a two-part system. Each system was made up of one satellite that would orbit Mars and a second piece that landed on its surface. Both missions lasted years beyond what was originally expected. The goal of the Viking missions was to test Mars's soil, study its atmosphere and weather, and perform biological experiments on the surface.

These biological experiments concluded what we may have suspected all along: There's no life on Mars.

MARTIAN MOONS

MARTIAN MOONS
BY THE NUMBERS

	PHOBOS	DEIMOS
DIMENSIONS:	17 × 14 × 11 MILES	9 × 7 × 6.8 MILES
DISTANCE FROM MARS:	3,700 MILES	14,573 MILES
LENGTH OF ROTATION:	7 HOURS, 39 MINUTES	30.3 HOURS
LENGTH OF REVOLUTION AROUND MARS:	7 HOURS, 39 MINUTES	30.3 HOURS
DISCOVERY DATE:	AUGUST 17, 1877	AUGUST 12, 1877

As you know, Mars is named after the Roman god. His Greek counterpart is Ares. Both gods have the same role: god of war. In Greek mythology, Ares has twin sons: Phobos and Deimos. Their names, which have been chosen for Mars's moons, mean *fear* and *flight*, respectively.

Both Phobos and Deimos are in a tidal lock with Mars, so the same side of the moons are always facing their host planet. On the **Martian Moons: By the Numbers** chart, you see that their orbital and rotational periods are equal. To always face the host planet, a moon must rotate at the same rate as it moves around its host planet, just like Earth's moon rotates around Earth.

MARTIAN MOON DISCOVERY

In the early 1600s, it was known that Earth had one moon. In 1610, Galileo discovered Jupiter's four largest moons. Early in the 17th century, German astronomer Johannes Kepler first proposed that Mars may host two moons. Kepler suggested multiple moons might exist because of Mars's location between the Earth, with one moon, and Jupiter, with at least four.

In August of 1877, American astronomer Asaph Hall discovered both Martian moons after a long search. At one point, Asaph was ready to give up, but his wife's words of encouragement kept him going. The very night that she spoke those words, Hall discovered Deimos, the smaller of the two moons and one of the smallest in the solar system. Hall continued to look closer to Mars. Six days later he discovered Phobos, just 3,700 miles above the Martian surface. Phobos is closer to Mars than Deimos, and closer to its host planet than any other moon in the solar system.

MARTIAN MOON MYSTERY

The origin of Phobos and Deimos is unknown, but most scientists believe that they were asteroids knocked from their orbits and captured by Mars. They are the same size, shape, and composition of the asteroids found in the neighboring belt.

Both moons are similar in composition, color, and shape. They're made up of ice and carbon-rich rocks. Phobos and Deimos are both heavily cratered and have surfaces made mostly of dust and rock-sized particles. Because of their very dusty surfaces, they appear very dark in the sky. Unlike the Earth's moon, which is a sphere, Phobos and Deimos have shapes closer to lumpy potatoes.

PHOBOS

Phobos is orbiting Mars so close to its surface that if you were at the north or south pole of Mars, you wouldn't even be able to see the moon.

ALTITUDE: CRITICAL CONDITION

Phobos is on a crash course with Mars because its orbit is deteriorating. Phobos comes 1.8 centimeters closer to Mars every year. It will take 50 million years, but eventually Phobos will either collide with Mars or break into a pile of rubble and become a ring around the planet.

FEATURES OF PHOBOS

Phobos has a large crater on its side called Stickney. It's named after Asaph Hall's wife, Angeline Stickney Hall. The Stickney crater is six miles in diameter, which is a very large indent on this small moon. Astronomers believe this crater was caused by a collision that nearly destroyed Phobos.

Phobos's surface is also covered in grooves, most of which are parallel to one another and run outward from Stickney. One theory suggests that the grooves are evidence that the small moon is collapsing as it moves closer to Mars.

DEIMOS

Deimos is the smaller potato of the two moons but was the first discovered because of its distance from Mars. While Phobos is on a collision course with Mars, Deimos has the opposite problem. This tiny moon is inching away from its host planet. Deimos will one day escape the Martian gravity and fly off into the solar system.

OH SO SMALL!

Deimos is one of the smallest moons in the solar system. As a result, it has very little gravity. Deimos's gravity is so weak that when the moon is hit by a meteoroid and dust flies up, the dust never resettles. It's ejected from the small moon and never comes back.

CRATERS

Deimos has two named craters on its surface, Swift and Voltaire, named after famous writers. More than 100 years before the moons' discoveries, both writers penned novels that included descriptions of Mars having two moons. The craters on Deimos's surface were named in their honor.

VISITING PHOBOS AND DEIMOS

While no spacecraft's primary mission has been to Phobos and Deimos, the Mariner 9 and Viking missions were the first to photograph the moons. From these photographs we learned about the characteristics of the moons' surfaces, as well as discovered their potato shapes.

ASTEROID BELT

You may have seen space adventure movies with scenes of starships speeding through the asteroid belt, ducking and weaving to avoid the giant rocklike objects. In reality, there's no need to duck and weave: The average distance between asteroids is more than half a million miles!

THE DISCOVERY

In the early 19th century, astronomers had been searching between the orbits of Mars and Jupiter for a missing planet. On January 1, 1801, Sicilian monk Giuseppe Piazzi announced the discovery of a body in the asteroid belt that is now considered a dwarf planet. Piazzi named the body Ceres, after the Roman goddess of harvest.

Fifteen months after the discovery of Ceres, on March 28, 1802, German astronomer and physician Wilhelm Olbers discovered Pallas, now the third largest known asteroid. In the years that followed, astronomers continued to find objects in the same general area of the solar system. By the 1850s, they began to refer to this region as the *asteroid belt*.

DEFINING THE REGION

The asteroid belt, also known as the main belt, is a disc-shaped region found between 2.2 and 3.2 AU from the Sun, between the orbits of Mars and Jupiter. The reason it's referred to as the *main belt* is because there are groups of asteroids in other regions of the solar system. Most asteroids, however, are found in this belt.

WHAT'S AN ASTEROID?

An asteroid is an irregularly shaped body made of rock and metal. Asteroids are classified into three types—C-type, M-type, and S-type—according to what they're made of.

C-type: The most common type of asteroid is the C-type, also called *chondrite asteroids*. They're made of clay and minerals called **silicates** and are very dark in color. Scientists believe that the Martian moons, Phobos and Deimos, were previously C-type asteroids.

M-type: M-type, or *metal asteroids*, are believed to be made mostly of the metals iron and nickel, the same metals found in Earth's core.

S-type: S-type, or *stony asteroids*, are believed to be a combination of the other two types; that is, a combination of both silicates and metals.

HOW MANY? HOW BIG? AND WHERE DID THEY COME FROM?

There are millions of asteroids in the main belt. Astronomers believe there are between one million and two million asteroids in the main belt with a diameter greater than a half mile. Most asteroids, however, are the size of a small rock, and can be as small as a dust-sized particle. The four largest objects in the main belt are Ceres, Vesta, Pallas, and Hygiea. Ceres, the largest, is one-third of the mass of all the objects in the main belt combined.

JUPITER, THE BULLY

Scientists believe that asteroids in the main belt might have held together to form a planet, if it weren't for Jupiter. Because Jupiter has the most mass of any planet, it also has the greatest gravitational force. Sometimes, if an asteroid comes too close to Jupiter, it can get pulled out of its orbit by Jupiter's gravity. This can create some havoc and give the asteroid a whole new **trajectory**, or path. Asteroids can be redirected right across the solar system by Jupiter.

WHAT ARE SILICATES?

The word *silica* or *silicates* refers to minerals or rocks that contain silicon dioxide (SiO_2). Yellow sand is made of silicates. Saying a rock or a mineral is a silicate just means that it contains SiO_2. Quartz, mica, feldspar, amphibole, olivine, and pyroxene are common minerals that are all silicates. Any rocks that contain these minerals, such as sandstone or granite, are silicates.

Earth's crust contains silicates. In fact, 90 percent of Earth's crust is silicon and oxygen. The reason Earth's crust has so many silicates is because when Earth cooled, the silicates were the least dense, and so they rose to the surface.

CERES

BY THE NUMBERS

DIAMETER: 588 MILES

DISTANCE FROM SUN: 2.8 AU

KNOWN MOONS: 0

LENGTH OF DAY: 9 HOURS, 4 MINUTES

LENGTH OF YEAR: 4.6 EARTH YEARS

AVERAGE TEMPERATURE: −37°F

CERES

Giuseppe Piazzi was looking for a star when he discovered Ceres. At first, Piazzi believed it was a comet, and for a short time, it was classified as one. It was later classified as a planet, and then as an asteroid. To this day, scientists aren't sure if Ceres originated in the asteroid belt or if it was pulled in from elsewhere in space.

In 2006, Ceres was upgraded to the status of dwarf planet at the same time Pluto was demoted from planet to dwarf planet. Ceres is the smallest of the dwarf planets, about the size of Texas. It's the only dwarf planet without moons. Ceres also lacks any rings and has a very thin atmosphere.

FROM ASTEROID TO DWARF PLANET

There are two main reasons Ceres was the only asteroid in the main belt to be upgraded to dwarf planet status. The first reason is its size. Ceres is the largest object in the main belt, and it makes up one-third of the main belt's mass.

The second reason is *because* of Ceres's mass. Ceres has a higher mass than other asteroids. This allowed it to pull loose sediment and solid rock closer to its center, which in turn evened out its surface, making it a sphere. The other large asteroids in the main belt are misshapen because they don't have the mass for this to occur.

FEATURES

Ceres is located right in between the inner and outer planets, but it shares more characteristics with the inner planets. It's small, with a rocky interior, no moons, and has no rings. Ceres has a weak atmosphere, composed primarily of water vapor.

Ceres is much less dense than the inner planets. This is believed to be due to its large content of ice. Ceres is believed to have a large amount of ice beneath its surface that could make up as much as one-fourth of the dwarf planet.

The surface of Ceres reveals some very reflective spots. Many of these brighter areas are found within craters. Astronomers believe they contain salts, and that the salts are remnants of waters that once existed on the surface. The water on Ceres has moved below the surface, leaving the salt to settle in the craters.

CERES'S ATMOSPHERE

As mentioned above, Ceres has a thin atmosphere made up of water vapor. This atmosphere appears and disappears according to the Sun, but not for the reason you might think. When objects from the outer atmosphere come closer to the Sun, their ices **sublimate**, or change directly from a solid to a vapor. Then, they refreeze when they're far from the Sun.

However, this is *not* what happens on Ceres. Instead, the ices on Ceres are sublimated when there is a large amount of solar activity. For example, when large amounts of particles are ejected from the Sun and strike Ceres, the energy they carry is transferred to the ice and it turns into a gas, resulting in Ceres's thin atmosphere.

VISITING CERES

In 2007, NASA launched Dawn, the first spacecraft sent to orbit, photograph, and study objects in the asteroid belt. Dawn first visited Vesta, and then Ceres.

VESTA

BY THE NUMBERS

DIAMETER: 326 MILES
DISTANCE FROM SUN: 2.36 AU
KNOWN MOONS: 0
LENGTH OF DAY: 5.3 HOURS
LENGTH OF YEAR: 3.6 EARTH YEARS
AVERAGE TEMPERATURE: –190°F

VESTA

Vesta is the second largest object in the asteroid belt after Ceres. It's extremely bright and makes up 9 percent of the mass of the asteroid belt.

HISTORICALLY SPEAKING

Now that Ceres is considered a dwarf planet, Vesta officially has the title of largest asteroid in the main belt. Wilhelm Olbers, who discovered the asteroid Pallas in March of 1802, discovered Vesta just five years later. Because Vesta was the fourth asteroid to be discovered, it was cataloged as, and is often referred to as, 4 Vesta. Olbers didn't name Vesta: He gave the honor to mathematician Carl Friedrich Gauss, who was able to calculate Vesta's orbit based on Olber's observations of the asteroid's movement. Gauss named Vesta after the Roman goddess of home and hearth, who was Ceres's sister.

FEATURES

Vesta, like other asteroids, has a flattened sphere shape. Vesta has no moons, no atmosphere, and no rings. Unlike other asteroids, Vesta appears to have layers much like Earth does. The core of Vesta is believed to be composed of iron and nickel, much like Earth's. Its core is surrounded by a rocky mantle, which is then covered in a crust made of basalt. Because of its layered body, Vesta is often referred to as a **protoplanet**, or the beginning of a planet.

Vesta is the brightest asteroid in the main belt. So bright, in fact, you can sometimes see it from Earth with the naked eye. Much like our Moon, Vesta has areas of varying brightness; some areas are darker, and others are brighter.

A GIANT CRATER

On the surface of Vesta, there's a crater that's 310 miles wide and 12 miles deep. The diameter of Vesta is only 326 miles, so the crater is almost as wide as the asteroid! This crater is believed to have been created by a large collision early during the formation of the solar system. The crater is referred to as the Rheasilvia Basin, and it's at Vesta's south pole. Vesta's surface also has a series of parallel grooves that run around the equator. These grooves are believed to have been created by this same collision.

The collision responsible for the Rheasilvia Basin may have also created additional asteroids in the process. These asteroids are collectively referred to as vestoids. Some vestoids are still in the asteroid belt, while others have fallen to Earth's surface as meteorites.

VISITING VESTA

Vesta was the very first asteroid to be orbited by a spacecraft. The Rheasilvia Basin was a point of interest to photograph as the spacecraft Dawn orbited Vesta. Dawn was able to confirm the size of the crater, as well as photograph and measure a large mountain at the bottom of the basin. The mountain measures twice the size of Mount Everest and is almost as tall as Olympus Mons on Mars. It's the second largest mountain in the entire solar system, and it's found on an asteroid!

VESTA VISITING

In the year 1960 in Millbillillie, Australia, two farmers observed a **meteoroid** fireball streak across the sky and crash to Earth. Ten years later, pieces from the crash were found. Upon studying the fragments, it was determined they came from the asteroid Vesta. Other **meteorites** found on Earth are also believed to have come from this asteroid.

ASTEROID, METEOROID, METEOR, METEORITE

An asteroid is a large, irregularly shaped rock that orbits the Sun and is too small to be a planet.

A meteoroid is a small piece of an asteroid or comet that is made of rock and found in the solar system. Sometimes, a meteoroid enters Earth's atmosphere. Many meteoroids burn up as they travel through the atmosphere. This happens because they're traveling very quickly and when they begin to hit air molecules, the air molecules provide friction that tears them completely apart. This usually results in a streak of light across the sky, or what we call a shooting star. Once this happens to a meteoroid, we call it a **meteor**.

Some meteoroids do not completely burn up, and a small piece may make it to the surface of Earth. Pieces of meteoroids that are found on the Earth's surface are called meteorites. To recap: An asteroid is a large rock in space, a meteoroid is a small one, a meteor is a meteoroid that has burned up in our atmosphere producing a streak of light in the sky, and a meteorite is a small piece of a meteoroid that has made it to Earth's surface.

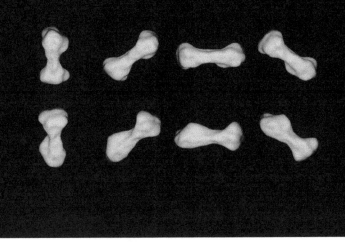

KLEOPATRA
BY THE NUMBERS

DIMENSIONS: 135 × 58 MILES

DISTANCE FROM SUN: 3.4 AU

KNOWN MOONS: 2

LENGTH OF DAY: 5.4 HOURS

LENGTH OF YEAR: 4 YEARS, 8 MONTHS

KLEOPATRA

Kleopatra is another prominent asteroid found in the main belt. It was discovered on April 10, 1880, by Austrian astronomer Johann Palisa and is named after the famous Egyptian queen. It's roughly the same size as the state of New Jersey.

Kleopatra is one of the more oddly shaped bodies in the solar system, and it has the nickname "dog bone asteroid" because of its two wide ends connected by a thinner region. Astronomers believe the shape signifies that is it the remnant of an ancient catastrophic collision. They have long puzzled over this asteroid's peculiar shape and the collision that could have caused it.

FEATURES

Kleopatra is an M-type asteroid. To map out its odd shape, astronomers bounced radar signals off the asteroid. It reflected the wavelength extremely well, which suggests it's likely made of iron and nickel. Its color also suggests it's made of metal. Because Kleopatra is so small, it doesn't have enough gravitational force to morph into a sphere. It has no atmosphere.

A DAY ON KLEOPATRA

Kleopatra rotates once every 5.4 hours. The axis of the oddly shaped asteroid runs perpendicular to the two large ends. This means that the two large ends are spinning around the central portion. In other words, it spins like a helicopter's blades.

TWIN MOONS

Kleopatra has two moons: Alexhelios and Cleoselene. They were both discovered in 2008 and named after Queen Cleopatra's twin children: Alexander Helios and Cleopatra Selene. Both moons are less than five miles in diameter. Having two moons means that Kleopatra is a triple asteroid system. In other words, the two moons and asteroid are all moving together.

It's likely that whatever event caused the formation of Kleopatra also created its moons. The moons are pieces of Kleopatra that remained close and were then pulled in by Kleopatra's weak gravitational field.

THINGS ARE FALLING APART

Once the moons were discovered, scientists were able to calculate the density of the asteroid. The density was surprisingly low. Because scientists already knew the asteroid was made of metal, they were able to calculate that 30 to 50 percent of the asteroid was empty space! The asteroid itself is not a solid body, but rather a pile of loose rubble held together by Kleopatra's gravity. This rubble will continue to travel in this form unless it's disrupted by an outside force.

ITOKAWA
BY THE NUMBERS

DIAMETER: 0.19 MILES
DISTANCE FROM SUN: 1.7 AU
KNOWN MOONS: 0
LENGTH OF DAY: 12.1 HOURS
LENGTH OF YEAR: 1.5 EARTH YEARS

ITOKAWA

Not all asteroids are found in the main belt. There are two other groups of asteroids in our solar system. A second group of asteroids are those found sharing the orbits of planets. These are called **trojans**.

A third group of asteroids are called near-Earth asteroids. These are asteroids that orbit the Sun, but their orbit also comes close to Earth's orbit. If the orbit of a near-Earth asteroid crosses over Earth's orbit, it is appropriately called an Earth-crosser.

Earth-crossers are broken into two groups: the Atens and the Apollos. These two groups of asteroids have different-size orbits. Itokawa is a member of the Apollo asteroids. There are more than 10,000 asteroids in the Apollo group.

FEATURES

Itokawa is asteroid number 25,143 and is an S-type asteroid. It's shaped like a peanut and is about the size of an oil tanker. Some portions of its surface are smooth, while others are rough. This is because Itokawa is actually a collection of loosely held pieces that are easily knocked loose when an object hits the asteroid. These pieces then resettle, resulting in some rough patches and other smooth ones.

Astronomers theorize that Itokawa might be a **contact binary system**. In other words, it might be two bodies stuck together, spinning around each other. The word *contact* refers to the fact that the pieces are touching, *binary* means Itokawa is made of two pieces, and *system* refers to the fact that these pieces are working together.

Think about the last time you blew bubbles. When two bubbles run into each other in the air, they don't pop. Rather, they come together. If one of the bubbles was moving faster than the other, they might spin around together. Astronomers believe Itokawa is made of two smaller asteroids that are stuck together and spinning around, like bubbles.

HISTORICALLY SPEAKING

Itokawa was discovered on September 26, 1998, by a near-Earth asteroid research team in New Mexico. It is named after Hideo Itokawa, who is considered the father of Japanese rocketry.

VISITING ITOKAWA

In 2005, a Japanese spacecraft called Hayabusa landed on Itokawa and collected samples of the surface dust, making it the first asteroid from which humans were able to collect samples. It returned to Earth with these samples in 2010. After studying these dust samples, along with the photographs taken on the mission, scientists came to two main conclusions. First, they confirmed that Itokawa is an S-type asteroid. Second, they concluded that Itokawa is not one large asteroid, but is made up of smaller pieces.

CHAPTER 4

THE OUTER SOLAR SYSTEM

The outer solar system begins at Jupiter's orbit and extends to the reaches of Neptune. In this region of our solar system, we encounter the four **gas giants**, more than 200 moons, and rings of ice and dust, among many other amazing things!

While the inner planets are collectively referred to as the terrestrial planets, the outer planets are called the gas giants because they are made up of giant collections of gas. Gas giants all have many moons and rings, though not all their rings are visible. These planets spin quickly but have very long orbits. This gives them short days and long years.

Objects moving in and out of the solar system are bounced around and easily swayed by the strong gravitational fields of the gas giants. Among the outer planets we find collections of ice, dust, and rock. Some objects are broken into pieces, forming new rings around planets.

Because the gravity of such large planets is strong, they've captured many moons. The planets' gravity has taken control of the moons' movement, putting the moons in orbit around each of them. Moons vary greatly in characteristics: some are dark, others are light; some are covered in ice, others are volcanically active; some are spheres, others aren't. Some moons spray vapor into the solar system, creating rings around the planets. Other moons act like a sponge cleaning a countertop, picking up pieces of the rings as they travel. One moon in the outer solar system has a thick atmosphere and liquid on its surface—the only body in the solar system other than Earth with this characteristic!

In this chapter, we will explore the outer solar system. We'll learn how it's ruled by gas giants, and we'll dig deeper into the characteristics of the many moons of these planets. We'll also learn how comets are born in the outer solar system, while meteors die there. So, let's blast off even farther into space to explore what's beyond the asteroid belt.

JUPITER

BY THE NUMBERS

DIAMETER: 86,881 MILES

DISTANCE FROM SUN: 5.2 AU

KNOWN MOONS: 79

LENGTH OF DAY: 9 HOURS, 56 MINUTES

LENGTH OF YEAR: 12 EARTH YEARS

AVERAGE TEMPERATURE: −238°F

JUPITER

Jupiter, named after the king of all Roman gods, is the first of the outer planets. As a gas giant, it is a ball of mostly hydrogen and helium, and it spins faster than any other planet in our solar system. Though it has not yet been confirmed, it's believed to

have a rocky interior, possibly the same size as Earth, at its core.

Jupiter is the most massive of all the planets in our solar system. In fact, its mass is so great, it likely stopped a fifth inner planet from forming. Jupiter can fit all of the other planets inside of it with room to spare!

JUPITER DOES NOT ORBIT THE SUN

Did you ever hold hands with a friend and spin in a circle together? If the two of you are the same size, you'd have the same mass, and your center of gravity would be at the point exactly in the center of you both. In other words, if you were looking down on the two of you spinning, the point you are spinning around is somewhere between you and your friend. This is called the center of gravity.

Now, what if you held hands with a two-year-old, and did the same thing? Your mass is surely greater than that of a two-year-old. If you did this, you would be swinging the two-year-old around you—and maybe leaning back just a little as you spin. In this case, the center of gravity would probably fall closer to you because you have more mass.

In space, the center of gravity of two objects is called the **barycenter**. For most planets, like Earth, the barycenter between Earth and the Sun is still within the Sun. This is because of the Sun's sheer massiveness and the fact that 1,300,000 Earths could fit inside of it.

Because Jupiter is more massive than all the other planets put together, its barycenter with the Sun is not within the Sun. Instead, it is about 30,000 miles above the surface of the Sun. This is the only planet that does not technically revolve around the Sun. Instead, it revolves around the center of gravity, which is 30,000 miles from the surface of the Sun.

FEATURES

You know Jupiter is enormous, but let's put it into some perspective. Jupiter's diameter is more than 11 times that of Earth's. That means you could line up 11 Earths side by side, and they still would not be as wide as Jupiter. You could fit 1,300 Earths into Jupiter. Only 1,000 Jupiters would fit into the Sun, compared with 1.3 million Earths.

Jupiter is made mostly of hydrogen and helium, although it also contains water, methane, and ammonia. The stripes you see on its surface are made of rising and sinking gases of different compositions. These gases rise and fall because of temperature differences.

Because of its size, relative distance, and composition, Jupiter is bright in our night sky. In fact, Jupiter is the fourth brightest object in the solar system. The Sun is obviously the brightest, followed by our Moon because it's so close. Venus comes in third because the light reflects off its clouds of sulfuric acid. Jupiter's size and composition make it fourth.

A DAY ON JUPITER

Jupiter is not only the largest planet, it also spins the fastest. A day on Jupiter is only 9 hours, 56 minutes. Because Jupiter spins so fast and is made of gas, it flattens at its poles and bulges at its equator, like a slightly squashed ball.

While Jupiter has the fastest day in the solar system, it takes Jupiter about 12 Earth years to orbit the Sun. From Earth, it appears Jupiter is moving extremely slowly.

THE GREAT RED SPOT

The Great Red Spot on Jupiter is caused by a hurricane-like storm, with winds topping 200 miles per hour. This spot has been visible since astronomers first saw Jupiter about 400 years ago. The storm itself is larger than Earth!

JUPITER'S RINGS

Most people know about Saturn's rings, but did you know all the outer planets have rings? Jupiter has four of them, though they're not visible from Earth. The Voyager 1 spacecraft discovered Jupiter's rings in 1979. When Voyager 1 passed

Jupiter, it took images of the planet with the Sun in the background, and these photos revealed the rings. The rings are made of dust. With the sunlight behind them, the dust particles reflected the light, making them visible. It's just like when sunlight is coming through a window in your house and you can see dust floating in the air.

Jupiter's rings were created by its inner moons. These inner moons experience meteoroid impacts, which create the dust.

INNER MOONS

Jupiter's inner moons are collectively referred to as the Amalthea group. The Amalthea group is made up of Metis, Adrastea, Amalthea, and Thebe. These four small moons are closest to Jupiter and are responsible for creating Jupiter's rings.

GALILEAN MOONS

Just outside the inner moons, and inside the outer moons, are the Galilean moons. These are the four largest moons of Jupiter. As their name suggests, they were discovered by Galileo in 1610. These were the first moons discovered orbiting a planet other than Earth. The Galilean moons were an important discovery. Their existence helped prove that Earth was not the center of the solar system. These four moons—Io, Europa, Ganymede, and Callisto—are among the most well-known moons in the solar system, and we will look at each in detail later in this chapter.

OUTER MOONS

Jupiter has 79 moons, and it is almost certain that more will be discovered in the future. Four of them are the inner moons, four are the Galilean moons, and the remaining 71 are outer moons. The outer moons are very small and they have very elliptical orbits. Many are believed to be asteroids that were captured by Jupiter's large gravitational pull.

A JOVIAN DISCOVERY

In January of 1610, Galileo first observed the four largest of the Jovian moons, which are what we call Jupiter's moons. Simon Marius claimed to have observed the four moons as much as five weeks prior, but Marius failed to record his first observations and Galileo provided in-depth documentation. As a result, Galileo is often credited with this discovery, and he had the honor of having this group named after him.

Upon discovery, Galileo named this group of moons the Medician planets, in honor of the Medici family. The Medicis ruled the city of Florence, Italy, where Galileo lived. Galileo referred to each of the individual moons by number: I, II, III, and IV. The moons were numbered in order from closest to farthest from Jupiter.

Galileo's system was used for hundreds of years. In the 1800s, as more moons were discovered, some started to find Galileo's system confusing. At this point, the names originally suggested by Marius were adopted: Io, Europa, Ganymede, and Callisto.

BY THE NUMBERS

DIAMETER: 2,264 MILES

DISTANCE FROM JUPITER: 262,000 MILES

LENGTH OF ORBIT AROUND JUPITER: 42 HOURS

LENGTH OF ROTATION: 42 HOURS

AVERAGE TEMPERATURE: −202°F

IO

Io is the fifth moon from Jupiter, and the first of the Galilean moons. It's named after one of Zeus's sweethearts. Zeus is the Greek counterpart to Jupiter.

Io is about the size of Earth's moon and is tidally locked with Jupiter. Io's claim to fame is that it is the most volcanically active body in the entire solar system. Io has a beautiful, yellow color due to the sulfur compounds found there. Because of Jupiter's size and distance from Io, the planet takes up almost one-fourth of Io's sky.

Io is different from most moons in the outer solar system, which are usually made up of a lot of ice. Io is the only moon in the outer solar system that contains no ice, but rather has an iron core and a rocky crust.

COSMIC CONTEST

Io is caught in a gravitational contest: It's constantly being yanked in multiple directions at the same time. Jupiter's gravity pulls Io in one direction, while the gravity of Europa, Ganymede, and Callisto pull Io in three other directions.

Because all these bodies are moving past one another and rotating at the same time, the surface of Io moves in *tides*, just like Earth's oceans and rocks move as they're pulled by the Moon's gravity. On Io, the rocky surface rises and falls! As you can imagine, this movement must have some side effects, and one of them is heat. The heat generated by this movement needs to go somewhere.

VOLCANOES ON IO

Io has hundreds of active volcanoes. The heat generated within the moon as it gets pulled and distorted causes volcanic eruptions, and these are not small eruptions. In fact, the lava is forcefully ejected from the moon, almost 200 miles high! The magma is like the basaltic magma found in Hawaii. It has a high iron content. On Io, there are also sulfur compounds found mixed into the magma. When the volcanoes erupt, this sulfur enters the atmosphere.

IO'S ATMOSPHERE

Io has a very thin atmosphere that is generated by the volcanic eruptions that occur there. The small amount of gas that is ejected with the magma forms a temporary atmosphere of sulfur dioxide. When Io moves into Jupiter's shadow, the temperature drops and the atmosphere collapses, forming frost on its surface. This sulfur dioxide frost is what makes the moon yellow.

AURORAS

There is a lot of magnetic activity that occurs between Io and Jupiter. There are forces at work that we cannot see. Jupiter's enormous magnetic field sweeps around the massive planet as it spins.

Jupiter's sweeping magnetic field removes particles from Io. These particles are carried to Jupiter and they interact with Jupiter's upper atmosphere to create auroras, just like the auroras created on Earth when our atmosphere interacts with winds from the Sun. This current is made possible by ionized gas that comes from Io's volcanoes.

EUROPA
BY THE NUMBERS

DIAMETER: 1,940 MILES

DISTANCE FROM JUPITER: 414,000 MILES

LENGTH OF ORBIT AROUND JUPITER: 85 HOURS

LENGTH OF ROTATION: 85 HOURS

AVERAGE TEMPERATURE: −260°F

EUROPA

Europa, named after another one of Zeus's sweethearts, is the second of the Galilean moons.

It's the smallest of the four, and it's slightly smaller than Earth's moon. Overall, it's the sixth largest satellite in our solar system. Europa has an iron core and rocky mantle, but the rest of it is water and ice. It appears white, like a ball of ice, with giant red cracks running all over its surface.

The outermost crust of Europa is a sheet of ice that's 10 to 15 miles thick. Beneath that coating of ice, the moon is thought to have an ocean more than twice the volume of Earth's oceans. Europa is very smooth, lacking impact craters. Because it's so smooth and covered in ice, Europa is one of the most reflective objects in the solar system.

EUROPA'S SURFACE

As mentioned above, Europa's surface is a sheet of ice. The moon is very cold, so all the water at its surface has frozen over. There is more water below the ice. Because of its low density, over one-fourth of Europa is thought to be composed of water.

The red lines visible on Europa's surface are cracks in the ice. Two things had to happen for those red lines to appear. First, the ice on the surface had to be cracked. Like Io, Europa is pushed and pulled by the gravity of Jupiter. Did you ever peel a hard-boiled egg? The outside of the egg is hard. Before you peel it, you might roll it on a hard, flat surface to crack the shell. This is what happens to Europa. It has a hard outer shell made of ice. Jupiter's gravity distorts the inside shape, resulting in the cracking of the icy shell.

Second, something filled the cracks and turned them red. When the cracks are created, substances from underneath the ice can come to the surface. Astronomers believe that these substances—which likely include sulfur compounds and salts—combine with the water ice. They are then affected by radiation from the Sun, and this turns them the reddish color that we see.

EUROPA'S ATMOSPHERE

Europa has a thin atmosphere of oxygen. Recently, it was discovered that it may be venting large plumes of water vapor into space. This would mean the moon is **geologically active**. Something must be occurring on the interior of Europa for geysers of water vapor to be shooting out of the moon into space.

The water vapor geysers have not yet been confirmed, but scientists have images of plumes and have detected water vapor close to Europa's surface. NASA hopes to launch a mission that will take samples from one of these plumes. This mission will allow them to confirm the composition of what they're seeing.

EUROPA CLIPPER MISSION

The Europa Clipper is set to launch in the year 2024. The Europa Clipper will fly by Europa 45 times, with each flyby occurring over a different part of the moon. It will be carrying various scientific instruments that will measure the temperature and composition of different areas on the moon. One of the reasons astronomers are interested in Europa and other bodies that harbor water is the search for life. The Clipper will also fly by Ganymede and Callisto.

GEOLOGICALLY ACTIVE VS. GEOLOGICALLY DEAD

Earth is a geologically active planet. Heat in the center of Earth is trying to escape. As it makes its way through the Earth's layers, it moves things around. This movement is visible in the form of volcanoes and earthquakes. Because things are moving around inside Earth, we say it's geologically active.

Some bodies in the solar system are not geologically active, but rather geologically dead. This means that things are *not* moving around inside. Everything has settled. No heat is trying to escape. There are no volcanoes, no quakes . . . there's simply no movement inside the body. Such a body can still have an atmosphere, and it can still have seasons. But, geologically, it is done transferring heat.

Three of the Galilean moons: Io, Europa, and Ganymede, are geologically active, but Callisto is geologically dead.

GANYMEDE

BY THE NUMBERS

DIAMETER: 3,274 MILES

DISTANCE FROM JUPITER: 665,000 MILES

LENGTH OF ORBIT AROUND JUPITER: 7.1 EARTH DAYS

LENGTH OF ROTATION: 7.1 EARTH DAYS

AVERAGE TEMPERATURE: –297°F

GANYMEDE

In Greek mythology, Ganymede was the cupbearer—the person responsible for pouring wine—of the Olympian gods.

Ganymede is the largest moon in the solar system. If Ganymede orbited the Sun instead of Jupiter, it would easily qualify as a planet. It wouldn't even be the smallest one! Ganymede is larger than Mercury and Pluto. Our Moon, the fifth largest satellite, is about two-thirds the size of Ganymede.

Like the other Galilean moons, Ganymede is tidally locked with Jupiter, so the same side of Ganymede always faces its host planet. This large moon rotates and completes one orbit around Jupiter in just longer than a week.

FEATURES

Ganymede has a magnetic field, which scientists believe is generated by its iron core. Surrounding the iron core are layers of silicate rocks, an ocean of salt water, and a 500-mile-thick crust of ice. The crust is also thought to contain a large amount of rocks. Ganymede's atmosphere is thin and made up of oxygen.

GANYMEDE'S SURFACE

Ganymede looks very similar to Earth's moon. Its surface has lighter and darker regions and multiple impact craters. One of these dark impact craters takes up a large portion of the moon.

Half of Ganymede is dark in color and covered in craters. These regions are older. The craters were created by impacts. The other half of Ganymede is lighter in color and covered in grooves. These grooves are the result of the surface being moved by Jupiter.

Ganymede is close enough to Jupiter that its surface is pushed and pulled in tides, just like the surfaces of Io and Europa. As a result, the crust has broken open. These cracks were pushed up as much as 2,000 feet, creating grooves. The grooves extend for thousands of miles. Along the grooves, water was released from underneath the ice. This created regions that are lighter in color.

There's a large dark area (the largest of its kind on Ganymede) named *Galileo Regio*, or the Galileo Region. Galileo Regio looks like a large dark circle surrounded by lighter areas on the moon's northern hemisphere. It has a diameter of almost 2,000 miles, and is close in size to the United States.

GANYMEDE'S MAGNETOSPHERE

The entire magnetosphere surrounding Ganymede is entirely within Jupiter's magnetosphere. These two magnetic fields have weird interactions with each other, causing auroras on both Ganymede and Jupiter. Io and Europa also cause auroras on Jupiter, even though those two moons do not have magnetic fields. Scientists are studying these interactions to help them learn more about Jupiter and its moons.

CALLISTO
BY THE NUMBERS

DIAMETER: 2,995 MILES

DISTANCE FROM JUPITER: 1,168,000 MILES

LENGTH OF ORBIT AROUND JUPITER: 16.7 EARTH DAYS

LENGTH OF ROTATION: 16.7 EARTH DAYS

AVERAGE TEMPERATURE: −218°F

CALLISTO

Like Io and Europa, Callisto, the fourth of the Galilean Moons, is named after still *another* of Zeus's sweethearts. Callisto appears to have a very dark body, with bright white speckles all over it. These speckles are craters. Callisto is the most heavily cratered moon in the solar system.

More than one million miles above Jupiter, Callisto is the third largest moon in the solar system, after Ganymede and Titan. Callisto is just slightly smaller than Mercury but has only one-third of Mercury's mass. This is because Mercury and Callisto have different compositions.

FEATURES

Callisto has a high iron content. It's 60 percent rock, with the remaining 40 percent water and ice. It also has the lowest **albedo**, or reflectivity, which is why it appears so dark.

Callisto is the least dense of the Galilean moons. The fact that it has a low density tells us that a large portion of the planet is water. The low albedo tells us that the surface has a lot of rock mixed in with the ice. If the surface were pure ice, it would appear brighter because it would be more reflective of the Sun's light. Like Europa and Ganymede, it's believed that there's an ocean beneath Callisto's exterior of rock and ice.

Callisto has a thin atmosphere of carbon dioxide.

GEOLOGICALLY DEAD

Callisto is only subject to Jupiter's gravity, and is not substantially affected by the other Galilean moons. It is also farthest from Jupiter of the four large moons, so while it's tidally locked with Jupiter, it does not experience the planet's pushing and pulling nor, therefore, the rising and falling of its surface.

Callisto is believed to be geologically dead. It does not release water vapor into space. Its surface is not cracked. It does not have grooves. And finally, heat is not generated in its core. As a result of it being geologically dead, Callisto is considered one of the less interesting bodies in our solar system to study.

CRATERS

Callisto is heavily cratered due to impacts from comets and asteroids. Its largest crater is named Valhalla, after the mythological castle that houses dead Norse warriors. On Callisto, the crater Valhalla is just shy of 200 miles wide and has eight impact rings extending outward for almost 2,000 miles. This basin is about the size of Ohio. The rings surrounding the basin could fit almost the entire United States.

The basin of the crater is brighter than the surrounding surface. This is most likely due to water ice that either covered the impact zone or was revealed beneath it. Surrounding the crater are ridges. These ridges are like the ripples created on a pond when you throw in a rock, except frozen in time.

The smaller craters on Callisto are surrounded by bright ringlets. These bright circles are believed to have been created by the impact that created each crater. Freshwater ice splashed up from the impact and settled in all directions on top of the rock-covered surface. Ice is more reflective than rock so, as a result, these rings appear bright.

SATURN

BY THE NUMBERS

DIAMETER: 72,367 MILES

DISTANCE FROM SUN: 9.6 AU

KNOWN MOONS: 82

LENGTH OF DAY: 10 HOURS, 42 MINUTES

LENGTH OF YEAR: 29 YEARS

AVERAGE TEMPERATURE: −288°F

SATURN

Saturn, named after the Roman god of wealth, is the sixth planet from the Sun and the second largest in the solar system. Saturn is also the least dense planet. Its density is so low, in fact, it could float on water! When most people think of Saturn, they think of its rings.

THE MAKING OF A FLATTENED BALL

Saturn is a gas giant made mostly of hydrogen and helium. These are the two lightest elements. The high abundance of these elements is why Saturn has such a low density.

Saturn is the second largest planet after Jupiter and has the second shortest day after Jupiter. A day on Saturn lasts only 10 hours, 42 minutes. Like Jupiter and the other gas giants, Saturn spins very fast but moves slowly in its orbit. A year on Saturn is equivalent to 29 Earth years.

Saturn's density is so low, and it spins so fast, that gas is spun out toward its equator. This is just like what happens on the merry-go-round on the playground. If you looked down on a merry-go-round, you'd see that everything flies out to the edges as it spins. This is because objects tend to travel in a straight line. With gas flying out toward the equator, Saturn is flattened at its poles, giving it the appearance of a flattened ball. This happens on all the gas giants to some extent, but it's most noticeable on Saturn because its low density exaggerates the effect.

SATURN'S SURFACE

Saturn does not have a true surface. It really is just a big ball of gas. This doesn't mean you could fly through Saturn, however. Traveling down into Saturn would be like swimming down to the bottom of a pool, except it's much deeper than any pool you've ever been in. The pressure of the gases above you would be crushing, much stronger than the water pressure in a pool.

Although Saturn is mostly hydrogen and helium, its upper atmosphere has small amounts of water vapor, ammonia, and methane. As some gases are heated and rise, and other gases cool off and sink, a churning of gases occurs in the atmosphere. These gases moving around in the atmosphere give Saturn its light yellow color.

SATURN'S RINGS

Saturn is the most recognizable planet because of its large, reflective rings. No other planet has rings as clearly visible as Saturn's. Unlike Jupiter's rings, which are made of dust, Saturn's rings are made of varying sizes of rock and ice. Recent discoveries show that one of them is being created by water vapor streaming out of one of Saturn's moons!

The pieces of rock and ice that make up most of the rings come from asteroids and comets that broke up and were captured by Saturn, and they range in size from a grain of sand to a boulder. Because Saturn's rings contain large chunks of ice, and ice is reflective, they're very easy to see, although you need a telescope to view them from Earth. A telescope with a magnification of 50 times will allow you to see Saturn's rings as a separate structure of the planet and in good detail. But you can see the rings with any amateur telescope that has at least 20 times magnification.

The rings are about 3,200 feet (0.6 miles) thick, and they extend for about 175,000 miles from the planet. There are seven main rings, often referred to as rings A, B, C, D, E, F, and G, lettered according to the order of their discovery. Images from the Cassini spacecraft mission (see page 102) show that these seven rings are composed of hundreds of thinner rings. There are also gaps between rings. The largest gap, called the Cassini Division, is between rings A and B.

Saturn is tilted 27 degrees, much like Earth. Because of its tilt, Saturn experiences seasons. When Saturn's rings are edge-on as viewed from Earth, it appears that they have disappeared! If someone held up a piece of paper perfectly flat and stood far away from you, it would look like nothing was there. This is what happens with Saturn's rings. They appear to not exist for a short period, until the planet has moved more in its orbit for the rings to become visible again.

Galileo is credited with first observing Saturn's rings. He did so with a telescope he built that had a magnification of 20 times. At first observation, in 1610, Galileo believed Saturn had two large spheres, one on either side. Two years later, he was stumped when these spheres seemed to have disappeared. Galileo had observed the rings edge on, as described above. In 1616, the spheres seemed to have reappeared. This time, Galileo described them as handles on either side of the planet. Years later, in 1655, Dutch astronomer Christiaan Huygens would observe the rings with a telescope that had a magnification of 50 times, which he built himself. Although Galileo was the first to spot the rings, Huygens is credited with correctly identifying and describing the ring structure we know today.

SATURN SYSTEM

Saturn has 53 confirmed and 29 unconfirmed satellites. This gives Saturn a total of 82 moons, the most in the solar system! Although Saturn is part of our solar system, it's a system in and of itself!

Saturn's moons can be broken into groups based on their size, composition, and orbits. Titan is the biggest of its large, icy moons, which also include Mimas, Enceladus, Tethys, Rhea, Hyperion, and Iapetus.

Saturn has several moons within its ring system. Some of these moons are creating the rings, some are collecting pieces of dust and rock from the rings and leaving gaps behind, and others are shepherd moons helping the rings stay together.

Once you remove the large, icy moons and the moons within the rings, Saturn has three remaining groups of moons: the Inuit group, the Norse group, and the Gallic group. Each of these groups is found outside of Saturn's rings. They are grouped according to their orbits' tilt and shape. Many of these small and oddly shaped satellites are believed to have been asteroids or comets that were captured by Saturn's gravity as they tried to pass the gas giant. Saturn's strong gravity forced each of these into an orbit, making it a moon.

TITAN

DIAMETER: 3,200 MILES

DISTANCE FROM SATURN: 759,000 MILES

LENGTH OF ORBIT AROUND SATURN: 16 DAYS

LENGTH OF ROTATION: 16 DAYS

AVERAGE TEMPERATURE: −290°F

TITAN

The first of Saturn's moons that we'll visit is Titan. In Greek mythology, the Titans were a group of 12 offspring that battled the Olympians. Many of Saturn's moons are named after individual Titans, including Hyperion, Iapetus, Phoebe, Rhea, and Tethys.

Titan is the most Earthlike body in the solar system. It has a thick atmosphere, clouds, and seas of liquid. Titan is the second largest moon in our solar system. It's larger than the planet Mercury and about 50 percent larger than Earth's moon. Titan could have been a planet, but it's orbiting the wrong body. Titan is the first and only outer solar system body on which humans have landed a spacecraft.

HISTORICALLY SPEAKING

Galileo had discovered Jupiter's four largest moons in January 1610 using a telescope he made himself. Christiaan Huygens, a Dutch astronomer and physicist, was inspired by Galileo's improvements on the telescope. Huygens began to make his own lenses and eventually used them in his own model of a telescope. On March 25, 1655, he was the first astronomer to observe Titan, Saturn's largest moon. He originally named it *Saturni Luna*, Latin for Saturn's moon.

FEATURES

Titan is made mostly of rock and ice. On the surface, there are lakes and oceans of methane and ethane. This means that other than Earth, Titan is the only body in the solar system with liquid on its surface. The liquids on Titan evaporate, form clouds, and rain back down onto the moon, just like water does on Earth. While the thick atmosphere might suggest life is possible, astronomers agree that it is far too cold at this distance from the Sun for life to be found.

ATMOSPHERE

Titan is the only moon in the solar system with a thick atmosphere. Ninety-five percent nitrogen and 5 percent methane, Titan's atmosphere extends 370 miles above its surface. In comparison, Earth's atmosphere extends only about 30 miles above its surface. Titan's atmosphere is about 50 percent heavier than Earth's. The molecules in Titan's atmosphere react with ultraviolet rays from the Sun to create an orange haze around the moon. This smog-like haze is so thick, we can't see the moon's surface.

DESTINATION: SATURN SYSTEM

The Cassini-Huygens spacecraft (often referred to simply as Cassini) is named after Giovanni Cassini and Christiaan Huygens. Both Cassini and Huygens made discoveries associated with the gas giant and its moons, and this spacecraft's mission was focused on Saturn.

In October 1997, Cassini was launched from Cape Canaveral, Florida. It took seven years for the spacecraft to reach its destination. After arriving, Cassini orbited Saturn, studying the planet, its rings, and its moons. The Cassini spacecraft was able to observe storms on Saturn. New moons were discovered, and numerous features of both rings and other moons were photographed.

In 2005, the Huygens probe was released from the spacecraft and safely landed on Titan. This is the only time a spacecraft has ever landed on an outer solar system body! The Huygens probe transmitted data from Titan's surface back to Earth for more than an hour.

After orbiting Saturn for 13 years, Cassini was low on fuel. It plunged into Saturn's atmosphere and burned up. This was done on purpose to avoid a collision with any of the smaller moons.

ENCELADUS
BY THE NUMBERS

DIAMETER: 313 MILES

DISTANCE FROM SATURN:
148,000 MILES

LENGTH OF ORBIT AROUND SATURN: 32.9 HOURS

LENGTH OF ROTATION: 32.9 HOURS

AVERAGE TEMPERATURE: −330°F

ENCELADUS

Enceladus is named after one of the giants in Greek mythology, even though it's not a giant itself. Enceladus is the sixth largest moon of Saturn and its diameter is just 10 percent of the diameter of Titan. About the same size as the state of Wyoming, it's one of the brightest moons in our solar system and is the most reflective body in the solar system. Since its discovery in 1789 by British astronomer William Herschel, scientists have questioned why Enceladus is so bright, but recent discoveries have revealed the answer.

BRIGHT WHITE

Some of the spacecraft Cassini's greatest discoveries were made about Enceladus. It was long known that the moon was the brightest body in our solar system. Cassini revealed that Enceladus is covered in a sheet of ice that is constantly being renewed. The greater ice content an object in space has, the more reflective it is. Because Enceladus constantly has a new ice surface, the ice is very fresh and not filled with dust. This bright ice reflects light, making Enceladus more reflective than any other body.

FLYBY

As Cassini was orbiting Saturn to study its rings, it was also recording Saturn's magnetic field. The readings of the magnetic field beneath Enceladus were being disrupted. Astronomers didn't know why. In a later orbit, scientists flew the spacecraft beneath Enceladus to get a closer look.

As it turns out, the southern pole of Enceladus is covered in parallel cracks, which were unofficially nicknamed "tiger stripes." The cracks are believed to be the result of the moon experiencing **tidal forces** as it moves around Saturn, in the same way that Io, Europa, and Ganymede are cracked on their surfaces because of Jupiter's gravity.

What was even more exciting is what's coming out of the cracks. Enceladus has gigantic plumes of water ice that shoot out of the cracks at a speed of 800 miles per hour, hundreds of miles into space. Many of these ice crystals fall back down onto the moon, creating its smooth, reflective icy surface. Some of the plume, however, creates Saturn's E ring.

On a later orbit around Saturn, in 2008, Cassini collected samples from the plumes of Enceladus and analyzed them, sending the data back to Earth. These samples revealed that the plumes contain water vapor, carbon dioxide, ice crystals, methane, and salts.

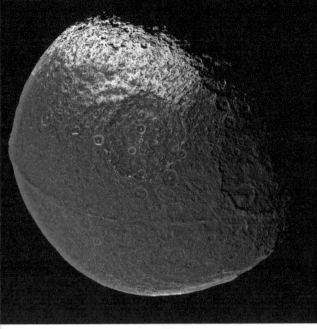

IAPETUS

BY THE NUMBERS

DIAMETER: 457 MILES

DISTANCE FROM SATURN: 2,213,000 MILES

LENGTH OF ORBIT AROUND SATURN:
79.3 EARTH DAYS

LENGTH OF ROTATION:
79.3 EARTH DAYS

AVERAGE TEMPERATURE: −261°F

IAPETUS

Iapetus, named after a Titan in Greek mythology, is the third largest of Saturn's moons and has some of the oddest lunar features of any moon in the solar system. More than two million miles away from its host planet, Iapetus is half black and half white. Despite its distance from Saturn, it's tidally locked, so the same side of it always faces Saturn. Iapetus has a mountain range that goes around its equator, making it look like a walnut!

HISTORICALLY SPEAKING

Iapetus was discovered on October 25, 1671, by Giovanni Cassini. Upon its discovery, Cassini noted that the moon was extremely visible during one part of its orbit, but not visible along the other portion of its orbit. The astronomer suggested that the moon must be tidally locked and consist of a brighter side and a darker side. Only these conditions could explain why the moon was only visible for part of its orbit. Cassini was right.

YIN AND YANG

One side of Iapetus is almost black, the other side is almost white. This earned it the nickname "yin and yang moon." Because of its extremely low density, we know that 75 percent of Iapetus is ice and 25 percent is rock. The white side is predominantly ice, making it very bright. The dark side contains more rock and dust, making it darker. But how did it get this way? There are several theories.

PHOEBE

Some astronomers believe that Iapetus began with an icy surface. Iapetus is far from Saturn, over two million miles. Even at this distance, Iapetus is the closest neighbor of Phoebe, which is another one of Saturn's small moons, over eight million miles from Saturn. While most Saturnian moons are icy worlds with high reflectiveness, Phoebe is dark in color and thought to have originally been a comet or other object from farther out in the solar system. Some scientists believe that Iapetus was once a bright, icy moon, but the debris that it collected from Phoebe darkened one side.

LAVA FLOWS

A second theory also begins with Iapetus as an ice world. This time, however, the darkening mechanism is the moon itself. Some believe ancient lava flows covered one side of the moon, creating the darker half.

HOT AND COLD

A third theory is referred to as *thermal segregation*. Thermal segregation means that one side is hot, and the other side is cooler. Some scientists believe that on one side of the moon, heat from the Sun caused the ices to evaporate, leaving behind darker material. This darker material then heated up faster, making more ices evaporate. The cycle continued to feed itself until one side of the moon was black and the other side remained white.

THE SURFACE

The darker regions on Iapetus's surface are smooth. They don't have many craters. For this reason, scientists believe the darkening agent is ongoing or at least very recent. It could be that the dark side was caused by a combination of things. For example, material from Phoebe may have landed on the moon, causing it to heat up and the ice to evaporate. The largest dark area is named *Cassini Regio*, or the Cassini Region, after the moon's discoverer.

EQUATORIAL RIDGE

Iapetus has a mountain range, an eight-mile-high ridge, that goes around the moon at its equator. Two other Saturnian moons, Atlas and Pan, have the same feature.

There are several theories about how this ridge may have formed. One idea is that the small moon may have eaten one of Saturn's rings. Saturn's rings only extend about a quarter million miles from the planet, though, and Iapetus is more than 2 million miles from the planet. Another theory is that Iapetus was struck by an object, such as an asteroid or a comet, that then broke up, possibly formed a ring around Iapetus, and then collapsed to its surface. A third theory is that the moon spun faster when it formed, creating the bulge at the equator, and then its rotation slowed.

VISITING IAPETUS

In 2004, Cassini was the first spacecraft to visit Iapetus and revealed many features of the moon. Because of its odd features, Iapetus continues to be of great interest to astronomers.

MIMAS

BY THE NUMBERS

DIAMETER: 246 MILES

DISTANCE FROM SATURN: 115,000 MILES

LENGTH OF ORBIT AROUND SATURN:
22 HOURS, 36 MINUTES

LENGTH OF ROTATION: 22 HOURS, 36 MINUTES

AVERAGE TEMPERATURE: −328°F

MIMAS

Mimas, named after another giant from Greek mythology, is the closest moon to Saturn. Mimas is only 115,000 miles from Saturn's surface. This little ice-ball of a moon has been called a snowplow because it clears the ice between two of Saturn's rings. Perhaps its most exciting feature, Mimas has a giant crater that makes it look like the Death Star from *Star Wars*!

HISTORICALLY SPEAKING

Mimas was discovered on September 17, 1789, by English astronomer William Herschel. Even though it's named after a giant, Mimas is extremely small—about the size of West Virginia. It was difficult to learn much about it until space travel was possible. The Cassini spacecraft and Voyagers 1 and 2 each provided more detail about the small moon.

CRATER CONUNDRUM

Most of this tidally locked moon is covered in large craters. Its largest crater is named Herschel, after William Herschel, and is easily its most distinguishable feature. This is the crater that gives it the Death Star look. At 80 miles wide, it is almost one-third the width of the entire moon. The walls surrounding the edge of the crater are three miles high. The impact that created this crater almost destroyed the moon. Cracks on the opposite side of the moon may have been created by that same impact.

Most of the craters on Mimas are very large. The craters at its south pole are far smaller, about half the size of the craters on the rest of the moon. It appears the surface of only the south pole has been renewed, or covered up.

ANOTHER ENCELADUS?

If you recall from an earlier section of this chapter, Enceladus has "tiger stripes" on its southern pole. These parallel cracks have huge water ice geysers shooting water vapor into space. These cracks are believed to be the result of tidal stress on the moon. But Mimas experiences far more tidal stress than Enceladus, and Mimas's orbit is far more eccentric. So, if Enceladus has cracks, why doesn't Mimas?

IS MIMAS A SNOWPLOW?

Mimas is found between rings A and B, the two largest and most identifiable rings around Saturn. Between these two rings are two things: Mimas and the Cassini Gap. Did Mimas *eat* a ring? Some astronomers believe that Mimas has acted like a snowplow and collected all the ice in its path around Saturn, leaving a gap behind in its wake. If Mimas picked up a ring, that would explain the resurfacing of the southern pole, despite not having any cracks.

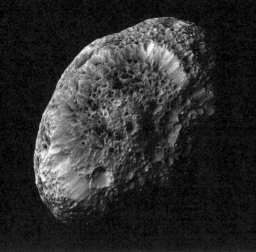

HYPERION
BY THE NUMBERS

DIAMETER: 168 MILES
DISTANCE FROM SATURN: 932,637 MILES
LENGTH OF ORBIT AROUND SATURN: 21 DAYS
AVERAGE TEMPERATURE: −292°F

HYPERION

Hyperion, named after a Titan in Greek mythology, is the largest nonspherical satellite in the entire solar system. Hyperion looks like an oddly shaped piece of dirty sponge traveling through space. It's absolutely riddled with holes and very dark in color. It's small, about the same size as the states of New Hampshire and Vermont combined.

HISTORICALLY SPEAKING

Hyperion was discovered by American astronomer William Cranch Bond and his son, George Phillips Bond, in 1848. English astronomer William Lassell discovered Hyperion at the same time, working independently from the Bonds. As a result, all three are credited with its discovery.

SPONGELIKE APPEARANCE

Hyperion has a crazy orbit. Its rotation is not predictable, either. It's essentially tumbling through space as though it's just been struck by something.

But Hyperion stands out most for its spongelike appearance. Certainly, the oddest feature when you look at Hyperion are all of its holes. How does something like that even happen? Scientists believe that this appearance is caused by **thermal erosion**, which begins with craters on the moon. Darker material collects within the craters. Because the material in the craters is darker, it heats up faster. As it heats up, it causes any ices beneath it to sublime, or evaporate. This deepens the crater. Large openings eat into the moon, leaving behind what looks like a sponge.

This small moon is believed to be composed of rock and ice, like the other moons of Saturn. The sides of Hyperion's holes, which are most likely made of ice, are bright.

COLORING

Hyperion is very dark in color. At the very least, it's coated in a darker material. Just beyond Hyperion's orbit is that of Iapetus, the "yin and yang moon" you read about earlier. If you remember, one theory concerning how one side of Iapetus was darkened was that debris from Phoebe, the next moon out, broke off and landed on Iapetus. It would only darken one side of Iapetus because Iapetus is tidally locked. Hyperion is *not* tidally locked. Its rotation is unpredictable. Some astronomers believe that Phoebe may also be responsible for coating Hyperion in this dark material.

URANUS

BY THE NUMBERS

DIAMETER: 31,518 MILES

DISTANCE FROM SUN: 19.8 AU

KNOWN MOONS: 27

LENGTH OF DAY: 17 HOURS, 14 MINUTES

LENGTH OF YEAR: 84 YEARS

AVERAGE TEMPERATURE: −353°F

URANUS

Uranus, named for the Greek god of the sky, was the first planet discovered by telescope. British astronomer William Herschel, also credited with discovering Enceladus, first saw Uranus on March 13, 1781.

The gas giants are much farther apart than the terrestrial planets. Uranus is 19 AU from the Sun, twice as far from the Sun as Saturn! Uranus has a beautiful blue methane atmosphere, spins on its side, and is just a bit larger than its ice twin, Neptune. Uranus is surrounded by 13 rings and 27 moons.

Although it was first discovered by telescope, Uranus is visible from Earth with the naked eye. With ideal conditions—clear weather, an extremely dark sky—you can see Uranus just by knowing where to look. If conditions are less than ideal, binoculars can do the trick. With the naked eye or binoculars, Uranus will look like a little dot or a star. But Uranus will appear as a blue-green disc when viewed through a telescope that magnifies 200 times.

FEATURES

Uranus is four times the width of Earth and more than 14 times more massive. Its atmosphere is composed mostly of hydrogen and helium with some water, ammonia, and methane. The beautiful blue color comes from the methane.

Beneath its atmosphere, Uranus is different than Jupiter and Saturn. Although its body is still considered a gas giant, Uranus is believed to have a rocky core surrounded by a mantle of ices made of water, ammonia, or methane. This mantle is surrounded by its atmosphere. Uranus doesn't have a true surface, but like the other gas giants, you can't fly through it.

A TILT LIKE NO OTHER

With a tilt of 98 degrees, Uranus spins on its side. Like Venus, Uranus also spins in retrograde, or backward. Uranus is the only planet whose tilt is perpendicular to its orbit. Because of this tilt, the poles of Uranus each face the Sun for half its orbit—42 years straight! You would think this would make the pole the warmest region on the planet, but that's not true. Scientists aren't entirely sure why, but the equator is the warmest region of Uranus.

When one of the poles moves from darkness into sunlight, storms erupt. The drastic temperature change causes movement in the atmosphere. This movement is responsible for the winds in the upper atmosphere.

RINGS

Uranus's rings appear to be almost straight up and down. Because the planet is on its side and the rings go around the equator, they're also on their side. The rings around Uranus are wobbly, unlike any of the other gas giants' rings.

Uranus sports two sets of rings. Its nine inner rings are darker in color. The outer rings are brighter. Uranus's rings are not nearly as spectacular as Saturn's. They are darker and much narrower, making them more difficult to see from Earth.

Uranus's brightest ring is called the Epsilon ring. The moons Ophelia and Cordelia are on either side of it. They're believed to be **shepherd moons**: they keep the ring together, just like a shepherd dog keeps a herd of sheep together.

The rings around Uranus are younger than the planet itself. Astronomers believe a solar system body was broken into pieces by Uranus's gravity. The planet captured these pieces and they stuck around, eventually forming rings.

MOONS

Uranus has 27 moons, but none are named after any mythological beings. Instead, most are named after characters from Shakespeare's plays. Some are named after characters from the writings of poet Alexander Pope. No one knows who started this, but it is possible that John Herschel, son of William Herschel, is to blame. Although John never discovered any moons himself, he held important roles in the field of astronomy and named the moons of Saturn that his father discovered. Despite not knowing who started the tradition of naming moons after literary characters, it has continued as additional moons have been discovered.

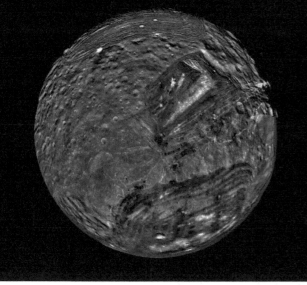

MIRANDA

BY THE NUMBERS

DIAMETER: 293 MILES

DISTANCE FROM URANUS: 80,641 MILES

LENGTH OF ORBIT AROUND URANUS:
1.41 EARTH DAYS

LENGTH OF ROTATION: 1.41 EARTH DAYS

AVERAGE TEMPERATURE: –335°F

MIRANDA

Uranus has five main moons: Miranda, Oberon, Titania, Umbriel, and Ariel. These are the only ones of Uranus's moons large enough to be rounded. Miranda is the smallest of these five, the closest to Uranus, and the last to be discovered. Oberon and Titania were discovered in 1787 by William Herschel, Umbriel and Ariel were discovered in 1851 by William Lassell, and Miranda was discovered by Gerard Kuiper on February 16, 1948.

For years it was believed that Uranus had only these five moons. In 1986, Voyager 2 passed Uranus and discovered an additional 10 moons, all of which were oddly shaped and much smaller than the main five. Since 1986, an additional 12 moons have been discovered orbiting Uranus, bringing the total to 27.

FEATURES

Miranda is only 293 miles wide, about the width of Pennsylvania. Miranda and the other moons of Uranus are all thought to be composed of about half ice and half silicate rocks. Since Miranda is not especially reflective, it's believed that its surface has experienced a large amount of cratering. Miranda has no atmosphere and no magnetic field.

FRANKENSTEIN MOON

Voyager 2 gave us our first ever look at the surface of Miranda. The images revealed the crazy, mismatched patches that earned it the unofficial nickname "Frankenstein moon."

These giant patches are referred to as *coronae*. A corona on a moon (or planet) is a circular or oval feature on the surface with grooves along the edges. Coronae are believed to be created by geologic activity. Hot magma within the mantle of the body rises to the surface, forming a bubble. Instead of erupting like a volcano, the magma seeps out of the sides of the bubble and collapses in the center. This creates ridgelike grooves surrounding the area where the bubble was. The only other body in the solar system with coronae is Venus.

The coronae on Miranda are approximately 120 miles wide, with depressions as deep as 12 miles. To give you an idea of how deep these are, compare them with the Grand Canyon, which has a depth of only one mile!

Miranda has three coronae, named Arden, Elsinore, and Inverness. Although not named for Shakespearean characters, they maintain the tradition by being named after locations in Shakespeare's plays.

VISITING MIRANDA

Very little is known of Uranus's moons because they are so far out in our solar system. The only mission that flew by Uranus and its moons was Voyager 2 in 1986.

TOP 12 LARGEST MOONS OF THE SOLAR SYSTEM

MOON	HOST PLANET
Ganymede	Jupiter
Titan	Saturn
Callisto	Jupiter
Io	Jupiter
Earth's Moon	Earth
Europa	Jupiter
Triton	Neptune
Titania	Uranus
Rhea	Saturn
Oberon	Uranus
Iapetus	Saturn
Charon	Pluto

DIAMETER: 30,599 MILES

DISTANCE FROM SUN: 30 AU

KNOWN MOONS: 14

LENGTH OF DAY: 16 HOURS, 6 MINUTES

LENGTH OF YEAR: 165 EARTH YEARS

AVERAGE TEMPERATURE: −353°F

NEPTUNE

Neptune, named after the Roman god of the sea, is the eighth planet from the Sun. At nearly three billion miles from the Sun, it's 30 times farther away than Earth is. Neptune is the smallest of the gas giants and is very similar in structure to Uranus. Neptune has five rings and 14 moons.

Neptune's claim to fame is the fact that it has exceedingly high wind speeds. In fact, it's the windiest body in the solar system.

DISCOVERING NEPTUNE

Neptune's discovery story is like no other. Almost every object in our solar system was discovered by either observation of the night sky or observation of a photograph. But not Neptune.

Neptune's discovery began with the study of Uranus. Astronomers noted that Uranus's orbit was off. Something beyond Uranus was affecting its orbit, pulling it out slightly farther from the Sun at different points. Using measurements of Uranus and its orbit, French astronomer and mathematician Urbain Le Verrier and British astronomer and mathematician John Couch Adams used mathematical calculations to find out what was disrupting Uranus's orbit. Each working independently in different locations, these men were able to calculate the mass and distance of an object beyond Uranus responsible for this disruption. In 1846, German astronomer Johann Gottfried Galle used these calculations to search for and find Neptune. All three are credited with the discovery.

FEATURES

Neptune is the smallest of the gas giants. Structurally, it's like Uranus, its ice twin. These gas giants are referred to as *ice twins* because both have a thick layer of ice as part of their core. Unlike Uranus, Neptune is tilted on an angle of 28 degrees, which is more similar to Earth's tilt. Neptune is thought to have a rocky core and a large amount of water, methane, and ammonia ices in its mantle. It has a thick atmosphere of hydrogen and helium, with traces of methane. Neptune's darker blue color is the result of the gas giant having more methane than Uranus.

Neptune's atmosphere has dark spots that appear and disappear on a regular basis. When NASA's Voyager 2 flew by Neptune in 1989, a storm named the Great Dark Spot was observed in the southern hemisphere. This storm was the size of Earth. Just five years later, the storm had disappeared and a new storm in the northern hemisphere, named the Northern Great Dark Spot, had appeared.

WILD WEATHER

Neptune is distant, cold, and windy. Winds of up to 1,500 miles per hour have been recorded on Neptune. To put that in perspective, think of a category 5 hurricane on Earth, which can have winds of 150 miles per hour. The winds on Neptune are about 10 times stronger than a category 5 hurricane!

Winds develop when there is uneven heating of air. Since Neptune does not receive much heat from the Sun because of its distance, astronomers believe it may have an internal heat source that creates this windy weather.

Neptune also has white, wispy cirrus clouds. These clouds appear and disappear quickly. All the clouds and storms on Neptune are in the outermost edges of its atmosphere.

GOT RINGS?

All the gas giants have rings, and Neptune is no exception. Neptune has five rings made of rock and dust, each a little more than a half-mile thick. It's believed that Neptune's rings were created by a moon that broke up. They are not very visible because they're made of rock and dust, so they do not reflect much light. Each ring is named after an astronomer who made a significant discovery associated with Neptune: Galle, Le Verrier, Lassell, Arago, and Adams.

VISITING NEPTUNE

Only one spacecraft has visited Neptune—NASA's Voyager 2. This spacecraft continued past Uranus in the late 1980s and photographed Neptune and its moons in 1989. When Voyager 2 flew past Neptune, it discovered all five rings and four moons. As previously mentioned, it also discovered the Great Dark Spot in the southern hemisphere, which has since disappeared.

TRITON

BY THE NUMBERS

DIAMETER: 1,682 MILES

DISTANCE FROM NEPTUNE: 220,438 MILES

LENGTH OF ORBIT AROUND NEPTUNE:
5.9 EARTH DAYS

AVERAGE TEMPERATURE: −391°F

TRITON

Triton is the largest of Neptune's 13 moons. Triton's diameter is about 80 percent of Earth's moon's diameter, less than two-thirds the width of the United States. It's the seventh largest moon in our solar system.

Triton is the son of Poseidon in Greek mythology. Poseidon is the Greek counterpart to Neptune, suggesting that Triton (the moon) is the son of Neptune (the planet).

Triton's orbit has a crazy tilt. This tilt creates seasons that last 41 years! (Now *that* would be a long winter.) Icy volcanoes in the southern hemisphere make Triton one of the few volcanically active moons in the solar system.

HISTORICALLY SPEAKING

Triton was discovered just 17 days after its host planet, Neptune, was discovered. It was found by William Lassell on October 10, 1846. From 1846 to 1949 it was simply referred to as "satellite of Neptune." Prior to 1949, it was the only known moon of Neptune.

FEATURES

Triton has a core made of silicate rock. Surrounding its core is an icy mantle, topped with a crust made of frozen nitrogen and methane.

Triton has a thin atmosphere made up of mostly nitrogen, with some methane. This atmosphere comes from volcanic activity. Triton is one of the few geologically active moons in the solar system.

In 1989, Voyager 2 captured images of five-mile-high plumes erupting from geysers located near Triton's southern pole. The plumes are shooting ice crystals made up of nitrogen and methane into the air, replenishing its thin atmosphere. When this thin atmosphere of nitrogen and methane freezes, it forms ice crystals on the moon's surface. The constant addition of layers of these ice crystals means that there are very few impact craters on Triton's surface. They are constantly being covered up. Just like falling snow would cover your footprints in the snow, Triton is constantly being resurfaced by these falling ice crystals. The layers of ice also make Triton reflective: It reflects 70 percent of the sunlight that reaches it.

A WONKY ORBIT

Triton is the only large moon in the solar system to orbit its host planet in the opposite direction that the planet is spinning. Both Venus and Uranus spin backward, which means they rotate in retrograde, but Triton *orbits* in retrograde.

Triton's axis is on a tilt of 30 degrees, which seems like no big deal. And it wouldn't be, except for the fact that Triton's orbit is also on a tilt of 157 degrees! Triton orbits Neptune almost perpendicularly. Triton's orbit leads it up above Neptune, and then down below Neptune, as Neptune orbits the Sun. This, combined with the fact that it takes Neptune 165 Earth years to orbit the Sun, leads to a very interesting situation for Triton. You see, for one-quarter of Neptune's 165-year orbit (41 years), Triton's north pole is directly facing the Sun. As Neptune moves around the Sun, Triton spends one-quarter of Neptune's orbit in each season.

CENTAURS

In Greek mythology, a centaur is a creature that has the top half of a man's body and the legs of a horse. In astronomy, centaurs have characteristics that blend different classifications of solar system objects, making the name *centaur* appropriate.

HISTORICALLY SPEAKING

In 1977, American astronomer Charles Kowal discovered Chiron. Chiron is a small body, about 125 miles in diameter, that orbits the Sun between Jupiter and Uranus. Its orbit is eccentric: Chiron crosses inside Saturn's orbit at perihelion and almost reaches Uranus at aphelion.

Chiron is made of rock and lots of ice. Initially, it was classified as an asteroid, the farthest out in our solar system, and assigned the number 2060.

Then, in 1989, a funny thing happened. Chiron was being observed when it happened to pass in front of a star. When an object passes in front of a star it gives astronomers a unique opportunity to view the object with light behind it. This can reveal new characteristics the astronomers may not have otherwise observed. When this happened, astronomers observed a haziness around Chiron. The fuzzy cloud around the asteroid suggested it may be more like a comet than an asteroid. Chiron was reclassified as a comet. This prompted the naming of Chiron. In Greek mythology, Chiron is a centaur: a combination of two things.

In the years that followed, more and more objects like Chiron were discovered. They orbited the Sun. Some have rings, like planets. Their orbits keep them within the outer solar system, between Jupiter and Neptune, and these orbits are eccentric. These objects are the size of asteroids, but sometimes they behave like comets. This group of objects did not fit well into any category, so a new classification of solar system objects was born: centaurs.

WHAT WE KNOW TODAY

By definition, a centaur is a small body that orbits the Sun between Jupiter and Neptune. Since the naming of the new classification, dozens more have been identified, and thousands more may be waiting to be discovered.

Centaurs are believed to be solar system bodies in transition. What does that mean? Centaurs are believed to come from the Kuiper Belt, which is an asteroid belt beyond Neptune (see page 133). Since it is much colder way out there, objects in the Kuiper Belt have more ice than objects in the asteroid belt and are often much larger. Neptune's gravity affects their orbit and sends them barreling toward the Sun. They regain *some* stability, orbiting the Sun on an eccentric path among the outer planets for a period. Their orbit never becomes balanced because they are crossing orbits of the gas giants.

As the outer planets play gravitational soccer with the centaurs, something is bound to give. The centaur can get kicked out of the stadium . . . that is, the solar system. Or Jupiter's massive gravity can slingshot it around the Sun, and the centaur will join Jupiter's family of short-period comets (see page 156).

We know the length of the centaur phase of an object's life is limited. This means that centaurs must be replenished. In other words, if we continually lost centaurs they would eventually be completely gone. But they're not gone, so the Kuiper Belt must be adding to them somehow.

DIFFERENCES

Since the discovery of centaurs, variations have been discovered within them. Some, for instance, have rings like the outer planets. And the colors of the centaurs can vary, from a bluish gray to a reddish hue, depending on their composition.

Saturn's moon Phoebe is believed to be a captured centaur. Phoebe is very small and dark, and it has a retrograde orbit around Saturn. It is very different from the other Saturnian moons and likely came from beyond Neptune.

In 2006, the first double centaur was discovered. A double centaur is two centaurs that orbit each other very close together, but also travel together in an orbit around the Sun.

ERIS

MAKEMAKE

HAUMEA

PLUTO

CHAPTER 5

WAY OUT BEYOND

Have you ever made a model of the solar system for school or from a kit? If so, it probably included the Sun, eight planets, maybe the asteroid belt, and possibly a few moons. But there is much more beyond those eight planets. If we scaled down the solar system so that an AU is about a centimeter, its radius would be about four and a half football fields. With the Sun at one end, Neptune would be just 30 centimeters away. That leaves over four football fields in all directions! That is a *lot* of space!

As it turns out, there are all kinds of icy bodies in this space. Just beyond Neptune's orbit, from 30 AU to about 50 AU from the Sun lies the Kuiper Belt. This is where we find Pluto and three other dwarf planets. Scattered disc objects orbit the Sun from beyond the Kuiper Belt. Much of this area of the solar system is still unexplored.

SPOILER ALERT #1: IT'S NOT ALL A DISC

Yes, the Sun and eight planets are found on the same plane. This causes all the planets, the Sun, and the constellations (also on the same plane) to follow the same path across our sky. But not all objects in the outer solar system follow this ecliptic path. This means that their orbit is not on the same plane as the orbits of Earth and the other planets.

SPOILER ALERT #2: SOME BODIES HAVE ECCENTRIC ORBITS

Astronomers spend a lot of time discussing the orbit of outer solar system bodies and comparing them with the orbits of the planets. The orbits of the eight planets are not perfect circles, but in general they have a low **eccentricity**, meaning their orbits are nearly circular. Many objects in the outer solar system, however, including Pluto, have eccentric orbits. This means that their orbits aren't circular, but rather are more of an elliptical shape. The eccentricity of orbits can be measured mathematically on a scale from zero to one. Zero is a perfect circle.

VOLATILES

The outer reaches of the solar system are often described as ice worlds. The farther from the Sun an object is found, the less heat there is. *Ice world* is not a surprising description. Many of the objects here contain frozen methane, water, carbon monoxide, ammonia, or nitrogen. These substances are referred to as **volatiles**. A volatile is a substance that easily goes from solid to a gas. That process is called sublimation, but the substance itself is referred to as a volatile. Volatiles will only sublimate when they pass closer to the Sun and they are heated. While objects found in the outer solar system contain volatiles, they are also composed of various rocks and dust particles.

WHAT'S IN A NAME?

Did you ever wonder who gets to name everything in space? Who's
in charge?

The **International Astronomical Union (IAU)** is the organization
that is tasked with selecting names. The IAU is made up of more than
13,000 members from more than 100 countries. Members suggest and
vote on names for dwarf planets, moons, and other bodies. There are rules
guiding the naming of some types of objects, such as planets and dwarf
planets. For example, dwarf planets are expected to be named after a
mythological creature associated with creation. Names have been taken
from many different mythologies, including Greek, Roman, Hawaiian,
Chinese, and Rapa Nui, to name a few.

KUIPER BELT

The Kuiper Belt is named after Gerard Kuiper, a Dutch scientist who, in 1951, first speculated about objects in our solar system beyond Pluto. The Kuiper Belt begins at Neptune's orbit, about 30 AU from the Sun. This region extends to about 50 AU from the Sun.

KUIPER BELT

While it's on the same plane as much of our solar system, the Kuiper Belt is more inflated than the rest of the disc. It is often described as having the shape of a donut or inner tube. Why? Because the objects travel on tilted orbits. They occupy a space that goes above and below the disc that is our main solar system.

WHAT IS THE KUIPER BELT?

The Kuiper Belt is a volume of space found beyond Neptune. Scientists believe there are hundreds of thousands of objects in the Kuiper Belt. So far, 2,000 of the objects have been tracked and recorded. While that number is impressive, the objects combined are believed to amount to less than 10 percent of Earth's mass.

The objects in the Kuiper Belt are made up of different types of ices (from water, methane, ammonia) and rock. Much of this matter is believed to be left over from the formation of the solar system. Scientists believe that Neptune's gravity is responsible for these objects having never formed a planet, just like Jupiter's gravity is responsible for the asteroid belt having never formed a planet.

There are three notable bodies in the Kuiper Belt region: Pluto, Haumea, and Makemake. These are three of the five dwarf planets in our solar system. We'll look more closely at each of these bodies in the sections that follow.

KUIPER BELT OBJECTS

A KBO sounds like something out of a science fiction book, but it's not. KBO stands for Kuiper Belt objects, also referred to as trans-Neptunian objects because they're found beyond Neptune. In this book, we will refer to them as KBOs.

In 1990, scientists began grouping KBOs. They quickly realized that they could group them according to their characteristics, including size, shape, and orbit. Names were given to the different groups of KBOs.

First, there are the classical KBOs. These KBOs are between 40 and 50 AUs from the Sun. All their orbits are mostly circular, meaning they have very low eccentricities. Some of these KBOs have their orbit on the same plane as the planets; this subgroup is called the cold classical KBOs and they're not affected by Neptune.

Some classical KBO orbits *have* been affected by the gravity of Neptune, which has essentially caused their orbit to be tilted in comparison with the planets. These are called the hot classical KBOs. This is one of the reasons the Kuiper Belt is a volume instead of an area. Many of the orbits rise above and then dip below the plane of the planets' orbits.

Another group of KBOs are called resonant KBOs. These are KBOs that resonate with Neptune. This means that there is a very precise rhythm to how they move in their orbit in relation to how Neptune moves in its orbit. To understand this idea, we will look at just one group of resonant KBOs: the plutinos.

One of the groups of resonant KBOs are called the plutinos. As you might have guessed, Pluto is a plutino. The group is named after the dwarf planet. The plutinos resonate at a ratio of 2:3 with Neptune. When we look at the ratios for resonance, the first number refers to the plutino and the second number refers to Neptune. A ratio of 2:3 means that for every two orbits the plutino completes, Neptune completes exactly three orbits. It's important to understand that these ratios are exact. If those ratios were off by even a tiny amount, their orbits would eventually have nothing to do with each other, and they would no longer resonate. This is like playing with a hula hoop. When you swing a hula hoop around your waist, your body resonates with the hoop. If the ratio of movements is off, even just a little bit, the hoop longer resonates, and drops to the ground.

The word *plutino* is often confused with the term *plutoid*. A plutoid is a dwarf planet beyond Neptune.

PLUTO

DIAMETER: 1,476.8 MILES

DISTANCE FROM SUN: 30 AU TO 50 AU

KNOWN MOONS: 5

LENGTH OF DAY: 153 HOURS

LENGTH OF YEAR: 248 EARTH YEARS

AVERAGE TEMPERATURE: −375°F

PLUTO

Pluto was once the ninth planet of our solar system. In 2006, it was demoted by the IAU to the status of dwarf planet. In July 2015, the New Horizons space probe sent the first up-close photographs of Pluto and its largest moon, Charon, back to Earth.

HISTORICALLY SPEAKING

Pluto was discovered in 1930 and initially classified as a planet. It was named by an 11-year-old girl named Venetia Burney, granddaughter of its discoverer, Clyde Tombaugh. She suggested the name Pluto after the Roman god of the underworld. In Roman mythology, the underworld is where souls go after a person dies. This seemed fitting as Pluto was extremely distant and cold. Pluto's moons are also named after characters from the underworld.

FEATURES

Because Pluto is so far from the Sun, it's very cold. The average temperature there is between −350 degrees and −400 degrees Fahrenheit. Pluto is made mostly of rock, but it also has ice, just like other KBOs. The ice is made of frozen water, nitrogen, methane, and carbon monoxide.

Pluto has an eccentric orbit that's also on a tilt. Pluto comes as close as 30 AU from the Sun and travels out to almost 50 AU from the Sun. Pluto is sometimes even closer to the Sun than Neptune. When Pluto travels closer to the Sun, some of the ice on its surface turns into gas. This makes Pluto's thin atmosphere temporarily thicker. When it travels farther from the Sun, those gases turn back into ice.

Pluto has no rings and is tilted on its side, like Uranus. Pluto has a retrograde rotation like Venus and Uranus. Unlike the surface of Mars, which is red because of the compound iron oxide, the red surface of Pluto is thought to be the result of a chemical reaction that occurs when sunlight interacts with its atmosphere.

HOW BIG IS PLUTO?

Pluto is sometimes referred to as "the King of the Kuiper Belt." It is the largest object in the Kuiper Belt. Although Pluto is the largest KBO, it's smaller than all the planets. In fact, it's even smaller than our Moon. If you could hold Pluto up to Earth, it would be about half the width of the United States.

WHY DID PLUTO GET DEMOTED?

We know that Pluto was originally classified as a planet and that it was demoted to a dwarf planet in 2006. But why? The discovery of Eris—a dwarf planet found in the scattered disc beyond the Kuiper Belt that's just slightly smaller than Pluto—in 2003 prompted this decision. Some scientists believed that Eris should be the 10th planet. Others disagreed. This dispute started a scientific debate, and in the end, scientists decided to clarify the definition of a planet. Part of this new definition requires that a planet have gravitational dominance and have cleared its path in its orbit. Neither Pluto nor Eris met these new criteria. Sorry, Pluto!

HEART SHAPE

One of the more recently discovered features of Pluto is the heart shape that appears on its surface. This region has been named the *Tombaugh Regio*, or the Tombaugh Region, named after the scientist who discovered Pluto, Clyde Tombaugh. The left side of the heart has a more defined appearance. This is a basin on the planet's surface. It is filled with frozen nitrogen. This nitrogen sublimates (goes directly from a solid to a gas) when Pluto is close to the Sun. Scientists liken this to Pluto's heart *beating*, because it causes winds in the atmosphere and possibly the movement of particles at the surface. The nitrogen sublimates, refreezes, sublimates, refreezes—you get the idea.

TRAVELING TO PLUTO

On January 19, 2006, the New Horizons space probe was launched from Cape Canaveral, Florida. It took more than nine years to reach its destination: Pluto. New Horizons was the first probe sent to photograph Pluto, its moons, and other objects in the Kuiper Belt. As of this book's publication, New Horizons is still in the Kuiper Belt, photographing other KBOs.

CHARON
BY THE NUMBERS

DIAMETER: 753 MILES
DISTANCE FROM PLUTO: 12,200 MILES
LENGTH OF ROTATION AROUND PLUTO:
6.4 EARTH DAYS
AVERAGE TEMPERATURE: –364°F

CHARON

Charon, one of Pluto's moons, is half the size of its host planet. No other moon in the solar system is half the size of the planet it revolves around. If Pluto is the same width as the United States, Charon is the width of Texas. The same side of Charon is always facing the same side of Pluto.

HISTORICALLY SPEAKING

In 1978, Charon was discovered accidentally by American astronomer James Christy. Christy was studying Pluto when he noticed that its shape looked wider than expected. After examining other photographs, he realized that what he was seeing wasn't just Pluto, but Pluto and a moon. Charon was the first of Pluto's five moons to be discovered. Charon is believed to have been formed by an early collision between Pluto and another KBO.

Because Pluto is named after the Roman god of the underworld, Christy suggested the name Charon for the moon. In Roman mythology, Charon is the ferryman who takes people across the river Styx (another one of Pluto's moons) to the underworld. In Roman mythology, Charon is pronounced *Karen*. Christy, however, is known to have pronounced it *Shar-on*. This was because his wife Charlene's nickname was *Char*. Christy had initially named the moon after his wife, and then discovered the happy coincidence that the Roman mythological figure's similar name fit the bill. Both pronunciations are used today.

DOUBLE PLANET SYSTEM

Pluto and its largest moon, Charon, are sometimes referred to as a double planet system. Why? Imagine for a moment that you have a Ping-Pong ball on a string. You can easily swing the Ping-Pong ball in a circle around your body. It is clear that you are more massive than the Ping-Pong ball, and that you are controlling the ball's movement around you. Your motion is not affected by the swinging of a tiny ball.

Now imagine you have very heavy bowling ball on a chain. In order to swing a bowling ball around your body you are forced to lean in the opposite direction and move in a coordinated motion. In this case, the mass of the bowling ball is enough to affect *your* movement. This is how Pluto and Charon move. Charon is about half the size of Pluto, so Pluto's motion is affected by Charon. They basically move around each other. This was one of the reasons Pluto was demoted from the status of planet to dwarf planet.

TIDAL LOCKING

When you think of tides, you probably think of the Moon's gravity pulling on the Earth's oceans, causing them to rise and fall approximately every 12 hours. And you would be right. We call those *tides*. But what is the tidal force? Well, this is slightly more complex than the gravity of one body affecting another. You see, while gravity depends on mass, it also depends on distance. So, when we look at the Moon's gravity having a stronger pull on the ocean closest to it, it also has the weaker pull on the ocean farthest from it. This is the tidal force. It is not just the stronger pull on something close to it. It's also the weaker pull on the *other* end of that very same object because that end is farther away.

Tidal locking is when it takes an object the exact same amount of time to rotate as it does to complete one revolution. Our moon is tidally locked to Earth. Because of this, we always see the same side of the Moon. Many of the larger moons in our solar system are tidally locked to their respective planets. Charon and Pluto are extra special because they are *both* tidally locked with each other. This means that the same sides of both objects are always facing each other.

HAUMEA

BY THE NUMBERS

DIAMETER: 770 MILES

DISTANCE FROM SUN: 35 TO 50 AUS

KNOWN MOONS: 2 (NAMAKA AND HI'IAKA)

LENGTH OF DAY: 4 HOURS

LENGTH OF YEAR: 285 EARTH YEARS

AVERAGE TEMPERATURE: −400°F

HAUMEA

Haumea, named after the Hawaiian goddess of fertility and childbirth, is a dwarf planet found beyond Neptune's orbit. It's shaped more like an egg than a ball, is spinning wildly fast, and is the most distant object in our solar system to have rings.

HISTORICALLY SPEAKING

Haumea was discovered on December 28, 2004. There is a bit of controversy over who should receive credit for this discovery, as two separate teams—one in the United States and one in Spain—claimed credit at the same time. Although 2004 is the official year of discovery, it was made based on data collected in 2003.

Haumea is a member of the plutoids. Like the other objects in the Kuiper Belt, Haumea is made mostly of rock and ice. Haumea is the fastest-spinning large object in the solar system. It completes a rotation in less than four hours. It is believed that Haumea was involved in a large-scale collision around the time the solar system formed. This collision is believed to be responsible for the formation of Haumea's two moons and its quick rotation. In 2005, scientists discovered its two moons, Namaka and Hi'iaka, named after Haumea's daughters. In 2008, the IAU officially named the dwarf planet and its two moons. These names were decided on after the moons were discovered in an observatory in Hawaii.

FEATURES

Haumea is long and narrow. It's shaped more like an egg than a sphere, probably due to its fast rotational speed. Haumea is believed to be made more of stone and less of ice, which is different than other KBOs. This is fitting for the namesake, as the mythological Haumea is also associated with the element of stone.

There is a red spot on Haumea's surface. Scientists believe it's a crater that may have been created by an impact. The spot gives us clues about the interior of the dwarf planet. Haumea was discovered just days after Christmas. When one of the teams observed this red spot, they decided to nickname Haumea "Santa."

In 2017, Haumea passed in front of a distant star. When this happened, the light of the star was blocked for longer than it should have been, based on Haumea's size. This led scientists to discover that Haumea has rings. Haumea is the only dwarf planet with rings and is the most distant object in our solar system with rings.

At 385 miles wide, Haumea is about the width of Colorado, but it's bright enough that it can be seen from Earth with a telescope. When solar system objects are bright, they are said to have a high albedo. Only stars generate light, so albedo is how much sunlight the object is reflecting.

Very little is known of Haumea's atmosphere, if it has any at all. The dwarf planet is believed to be mainly rock, with a crystalline ice coating of a surface. Its moons are believed to have the same composition.

No spacecraft have traveled near Haumea yet. Everything we know about this dwarf planet was discovered from ground telescopes.

MAKEMAKE
BY THE NUMBERS

DIAMETER: 888 MILES
DISTANCE FROM THE SUN: 46 AU
KNOWN MOONS: 1 (UNCONFIRMED)
LENGTH OF YEAR: 305 EARTH YEARS
AVERAGE TEMPERATURE: −406°F

MAKEMAKE

Makemake, pronounced *mah*-kee *mah*-kee, is two-thirds the size of Pluto. It's the second brightest object in the Kuiper Belt. Originally nicknamed the "Easter Bunny," Makemake was the fifth of the dwarf planets to be discovered.

MAKEMAKE'S DISCOVERY AND NAME

Makemake was discovered on March 31, 2005. It was nicknamed the "Easter Bunny" because the Easter holiday was just four days prior. The nickname was given by the same team that gave Haumea its nickname "Santa." That team, from Palomar Observatory in California, had also discovered Eris in October 2003 and was one of the teams that claimed the discovery of Haumea in December of 2004. All three of these dwarf planets were discovered in a span of just shy of 18 months, and their discoveries prompted Pluto's reclassification.

When the time came to officially name the "Easter Bunny," Makemake was chosen from the mythology of the Rapa Nui, the Polynesian people who live on the South Pacific island known as Easter Island. In Rapa Nui mythology, Makemake is the god of humanity and fertility.

FEATURES

Makemake is found in the Kuiper Belt, 46 times farther from the Sun than Earth is. While it takes sunlight about eight and a half minutes to reach Earth, it takes more than six hours for it to reach Makemake. This is why KBOs are often referred to as frozen ice worlds. While the length of the day is similar, it takes Makemake more than 300 Earth years to complete just one orbit around the Sun.

The surface of Makemake is believed to be similar to that of Pluto and the other KBOs. It has a reddish brown color, most likely due to the presence of frozen methane and ethane crystals. It's also believed to contain frozen nitrogen. In 2011, when Makemake passed in front of a star, astronomers were able to observe that it lacked an atmosphere. Scientists believe that Makemake might sublimate and refreeze a very thin atmosphere when it passes closer to the Sun, much like Pluto and other KBOs. The process of observing characteristics of a planet or dwarf planet when it passes in front of a star it is called *occultation*. Occultation doesn't allow us to see the object, but because there is a light behind it, we can tell if there is an atmosphere, or rings, or other objects around it.

In 2016, a satellite was seen orbiting Makemake. This moon has not yet been confirmed and is currently called MK2.

VISIBILITY

Makemake is the second brightest object in the Kuiper Belt, and is about two-thirds the size of Pluto.

It is believed that Clyde Tombaugh, who discovered Pluto, would have also discovered Makemake. However, at the time of Pluto's discovery, the positioning of Earth and Makemake would have placed Makemake in front of the Milky Way galaxy. This would have made it basically impossible for Tombaugh to pick it out among so many stars.

SCATTERED DISC

Where is the **scattered disc**? And *what* is the scattered disc? Well, this is one of those not-exactly-defined parts of astronomy that scientists are still discussing.

According to some, the scattered disc region overlaps with the outer edge of the Kuiper Belt. Others define it as a separate region. The Kuiper Belt ends at around 50 AU and the scattered disc is believed to extend as far as 1,000 AU from the Sun. One theory is that the icy minor planets that make up the scattered disc were flung into the outer regions of the solar system by the gas giants, early in the formation of the solar system.

Objects found in the scattered disc region are logically referred to as scattered disc objects, or SDOs. The Kuiper Belt and the scattered disc region transition the solar system from large planets to the outer Oort cloud. The Kuiper Belt is a more controlled area, while the scattered disc is more, well, scattered.

KBO VS. SDO

The objects of the Kuiper Belt and the objects of the scattered disc behave differently. KBOs have more regular orbits. That's to say, their orbits are closer to being circles. They're only slightly tilted in comparison with the solar system plane.

On the other hand, objects of the scattered disc are more erratic. SDOs orbit *way* above the plane of the other planets and then dip way below. Their orbits are tilted and highly eccentric, but are still influenced by Neptune's gravity, just like KBOs. Both KBOs and SDOs are sometimes referred to as trans-Neptunian objects because they are found beyond Neptune.

THE MOST WELL-KNOWN SDO

Eris is the most well-known and largest object of the scattered disc. Eris averages 68 AU from the Sun, far beyond the distant edge of the Kuiper Belt. Eris's orbit is on a tilt of 44 degrees. The orbit of Pluto, found in the Kuiper Belt, is on a much less drastic tilt, just 17 degrees. Many other objects have been found in the far reaches of the scattered disc. Newly discovered objects receive temporary names as astronomers continue to study them and determine how to officially classify them. Scientists believe that there are still many dwarf planets in this region that are waiting to be discovered.

ERIS

BY THE NUMBERS

DIAMETER: 722 MILES
DISTANCE FROM SUN: 68 AUS
KNOWN MOONS: 1 (DYSNOMIA)
LENGTH OF YEAR: 557 EARTH YEARS
AVERAGE TEMPERATURE: −400°F

ERIS

Either Eris was the 10th planet or Pluto was not a planet at all. Like Haumea and Makemake, Eris was discovered by a team of astronomers at the Palomar Observatory in California. Eris, originally nicknamed "Xena," was at first thought to be the 10th planet. At the time of its discovery, it was known that Xena was most likely larger than Pluto, three times farther from the Sun, and had a crazy orbit. Xena's orbit is tilted far out of the plane of all the other planets, has a high eccentricity, and doesn't have the Sun anywhere near its center. Xena also travels far beyond the reaches of the Kuiper Belt into the largely unknown scattered disc region.

Pluto is part of a binary system, has a slightly tilted orbit, and does not fit in with the outer planets. So, was Xena the 10th planet? If Pluto was a planet, surely Xena was, too. This is where the discussion began.

In the end, it was decided neither were planets. The definition of a planet was reestablished, and both Pluto and Xena, now officially called Eris, were classified as dwarf planets. Eris's existence was essentially the biggest argument for Pluto not being a planet. Mike Brown, the lead researcher at Palomar who discovered Eris, Haumea, and Makemake, even has the Twitter handle @plutokiller.

Ceres had also been discovered by this time and had been classified as an asteroid, the largest in the main belt between Mars and Jupiter. Ceres was also reclassified as a dwarf planet. Haumea and Makemake followed soon after.

A FITTING NAME

In Greek mythology, Eris is the goddess of chaos, discord, and strife. One of Eris's powers is to cause confusion and conflict. Fitting, as this was the discovery to cause debate among scientists. Eris's moon, Dysnomia, is named after Eris's daughter, the goddess of lawlessness. After Charon, Dysnomia is the second largest dwarf planet moon.

FEATURES

Eris is just slightly smaller than Pluto, both measuring about two-thirds the size of Earth's moon. Even though that seems small, all the objects within the asteroid belt could fit inside Eris.

Eris is mostly grayish white in color and is believed to be a mostly rocky body. Like other bodies from the icy regions of the solar system, it might generate a thin atmosphere when it travels closer to the Sun, only to have the atmosphere collapse as it travels back to the icier regions. Also like many objects from the outer solar system, these ices most likely include methane. The presence of methane ice crystals on its surface make Eris especially bright. Scientists speculate that ice might be renewed on its surface regularly, possibly from its interior.

Eris's orbit extends far beyond the Kuiper Belt into the scattered disc region of the solar system. Eris's orbit is tilted 44 degrees from the plane of the solar system. Because of this large orbit, it takes the dwarf planet 557 Earth years to orbit the Sun.

GONGGONG

Gonggong, a body about the size of Charon and very likely the sixth dwarf planet, is the first large solar system body to be given a Chinese name.

SNOW WHITE?

Gonggong was first observed in 2007. At the time, it was the largest unnamed object in the solar system. Megan Schwamb is credited with its discovery. Schwamb was a graduate student studying at Caltech, home of the Palomar Observatory. She, along with Mike Brown and David Rabinowitz, originally nicknamed the body "Snow White." They were expecting it to be a primarily white object. Much to their surprise, it turned out to be red!

NAMING GONGGONG

In 2019, Schwamb asked for the public's help in naming the body, and the name Gonggong won. Gonggong, the Chinese water god, has the head of a human with red hair, and the body of a serpent. In Chinese mythology, Gonggong causes destruction in the form of floods. The character is even credited with knocking Earth onto a tilt. Gonggong has one moon, Xiangliu. In Chinese mythology, Xiangliu is a nine-headed poisonous snake that also causes destruction.

WHAT DO WE KNOW SO FAR ABOUT GONGGONG AND XIANGLIU?

Gonggong is about half the diameter of Pluto (about the size of Charon) and is found in the scattered disc. Its orbit size is similar to Eris's, completing one orbit in approximately 554 Earth years, but it's on less of a tilt. Eris's orbit is tilted 44 degrees compared with the solar system, while Gonggong is tilted 30 degrees. Although it's red, it is believed that there is methane ice on its surface.

Gonggong is most likely a dwarf planet, but that has not yet been confirmed. As of now, it is just one of many other already identified probable dwarf planets, including DeeDee and Sedna.

COMETS

Comets are often referred to as "dirty snowballs." Comets are actually composed of a mix of ices, dust, and rock. Unlike KBOs, which are mostly rock with some ices on the surface, comets are an even mix of the two throughout. The actual solid comet made of ice, dust, and rock, is referred to as the nucleus. When the comet comes close to the Sun, the ices sublimate, or turn directly from a solid to a gas. This creates a fuzziness around the comet called the coma (*coma*: Greek for hair). This is where the word *comet* comes from. Comets, with the coma around them in the sky, looked like hairy stars to the ancient Greeks.

TAILS

Comets have two tails, an ion tail and a dust tail. One of the tails is created by the ice sublimating into gas. The other is created by the dust particles being released as a result of the ice sublimating.

ION TAIL

When the ice turns to a gas, it's superheated by the Sun. The gases released are charged particles, or ions. One of the gases released, for example, is carbon monoxide. When carbon monoxide is on Earth, it's not normally an ion. Because it's so bright near the Sun, however, one of carbon monoxide's electrons is stripped away, leaving it with a positive charge. Ions coming off a comet are easily blown by the solar wind from the Sun. For this reason, the ion tail is always pointed away from the Sun. Different gases appear different colors. For example, carbon monoxide creates a blue ion tail. Most comet ion tails are blue or green.

DUST TAIL

As the ice sublimates, pieces of dust loosen on the surface and leave the comet. This creates a dust tail. Dust tails are not affected by the solar wind like an ion tail. A dust tail is left behind a comet, marking its path. Because it initially leaves the comet with the gas, it can be slightly curved, but ultimately follows the comet. Dust reflects sunlight, making the dust tail white.

SHORT- AND LONG-PERIOD COMETS

Comets that complete one full orbit around the Sun in less than 200 years are called *short-period comets*. These comets originate in the Kuiper Belt and are generally found closer to the ecliptic. Their orbits tend to be on less of an angle. There are two families of short-period comets: Jupiter family comets and Halley family comets.

JUPITER FAMILY COMETS

Jupiter family comets have an orbital period of less than 20 years. Although they originate in the Kuiper Belt, they do not return there. A collision sends them inward, and Jupiter's gravity determines their ultimate final orbit. These comets have less volatiles, or ices that sublimate, because of their frequent trips around the Sun.

HALLEY FAMILY COMETS

Halley family comets have orbital periods of between 20 and 200 years. The most famous comet of this group is Halley's Comet, named for Englishman Edmond Halley, who discovered it. Halley's Comet completes an orbit around the Sun every 76 years. It was last visible from Earth in 1986.

Long-period comets are thought to come from the Oort cloud (see page 160). These comets take longer than 200 years to complete an orbit, and their orbits are highly inclined.

DIRTY SNOWBALL

The nucleus of a comet is compacted dust and ice. The surface of the nucleus is not shiny. This means that most of the surface is rocky or dusty, not icy, making it a bit like a dirty snowball. Photographs of comets show that their surface is rough and cratered. This means that the gases escaping are mixed in with the dust and rock. As a comet heats, the gases escape from inside of it.

CHAPTER 6

INTO THE UNKNOWN

The solar system is vast. It's so vast that astronomers don't know exactly where it ends. The hypothesized edge of our solar system is thought to be anywhere from 50,000 to 100,000 AU from the Sun. We've never seen this edge, but we have seen evidence of it in the form of comets.

OORT CLOUD

The Oort cloud is a hypothetical sphere of icy objects that surrounds our entire solar system. Its existence has not yet been proven. Astronomers believe there are anywhere from billions to possibly trillions of objects in the Oort cloud.

HISTORICALLY SPEAKING

The Oort cloud is named after Dutch astronomer Jan Hendrik Oort. Throughout the 20th century, Oort studied various astronomical bodies and formulated ideas. He first suggested our galaxy was spinning and helped calculate our solar system's location within the Milky Way. He also studied long-period comets, which led him to hypothesize the existence of the Oort cloud.

SIZE AND DISTANCE

The Oort cloud is *extremely* far away. Pluto is 39 AU from the Sun, on average. Eris is 68 AU from the Sun, about twice Pluto's distance. The Kuiper Belt is 30 to 50 AUs from the Sun. The Oort cloud *begins* between 2,000 and 5,000 AU from the Sun and extends to approximately 50,000 AU from the Sun. It surrounds the entire solar system—Kuiper Belt and all. It's *thousands* of times farther from the Sun than the Kuiper Belt.

The Oort cloud is believed to extend up to one-quarter of the way to the next closest star, Alpha Centauri. While Alpha Centauri is technically three stars, it appears as one in our sky. Two of the stars are a known binary system: Alpha Centauri A and B. A third star, Proxima Centauri, is thought to possibly be a third part of this system. That would make it a three-star system, but this is not yet confirmed.

Currently, Proxima Centauri is the star closest to our solar system, at 4.22 light years away. There are 63,241 AU in one light year (see page 163). If the Oort cloud extends to 50,000 AU, this is almost one-quarter of the way to Proxima Centauri.

FEATURES

The Oort cloud is believed to have two regions. The first region is the inner Oort cloud, or the Hills cloud. This inner region begins as more of a disc shape and transitions into the outer Oort cloud, which is a sphere.

OORT CLOUD VISITORS: LONG-PERIOD COMETS

Long-period comets have orbits that are all over the place. Because these comets come from every angle and take so long, it is believed that the Oort cloud is a sphere that is very far away, but still part of our solar system.

The Oort cloud exists at the hypothetical edge of where the Sun's gravity becomes so weak that the gravity of other bodies can steal comets away from the Sun. Remember, the gravitational force depends on the size of an object and the distance from an object. So, while the Sun is very massive, at this great distance its gravitational force would be very small. These objects are hanging on by a cosmic thread.

Astronomers believe that because the Sun's control over objects at this distance is so minimal, they are easily knocked out of orbit. How? The gravitational pull of other stars or the Milky Way galaxy itself can send objects from the Oort cloud shooting in toward the inner solar system or send them out of the solar system completely.

Long-period comets are of considerable interest for two reasons. The first is that they are rare. With orbits of more than 200 years, nobody on Earth will ever see the same long-period comet twice. The second reason is their brightness. Because of their infrequent trips around the Sun, they maintain a large portion of their ices. They have more to burn, with greater brightness, than comets that are constantly visiting the Sun.

HALE-BOPP

The most famous long-period comet is perhaps Comet Hale-Bopp, that last passed Earth in 1997. Comet Hale-Bopp was discovered by two amateur astronomers at the same time, each working separately. The comet is named after both of its discoverers, Alan Hale and Thomas Bopp. It takes Hale-Bopp 2,533 years to complete one orbit around the Sun. Don't bother marking its return on your calendar—it won't pass Earth again until the year 4385!

WHAT'S A LIGHT YEAR?

When we hear the unit *year*, we tend to think of a unit of time. But when talking about space, a *light year* is used to measure distance. More specifically, it's how far light travels in a year in the vacuum of space. Light is an electromagnetic wave. Nothing travels faster than light in a vacuum. Because space has such vast distances, we often describe them in light years. This is especially helpful when looking at the distances of stars.

The closest star to our solar system is Proxima Centauri, part of the Alpha Centauri system. This star is 4.22 light years away. This means that when you are looking at this star in the sky, you are seeing light that left the star 4.22 years ago!

The star Tau Ceti is just less than 12 light years away. If you are younger than 12 years old and you are looking at Tau Ceti in the night sky, you are viewing light that left the star before you were born!

In essence, when we look at stars in the sky, we are looking into the past. Pretty cool, huh?!

NEIGHBORING WORLDS

The scientists at the **SETI** Institute are looking for a habitable planet, much like Earth, among the other stars in our galaxy. (SETI is an acronym for the **s**earch for **e**xtra**t**errestrial **i**ntelligence.) They, and other astronomers, have found thousands of planets, but only a few that could be habitable. What they are looking for specifically is an Earth-sized planet with an ozone layer and liquid water, that is in what is called the **habitable zone**, or within a certain distance from its star. The planet closest to this description lies in the habitable zone of Proxima Centauri. Will we ever travel there? It is more than four light years away, so even if we could travel at the speed of light, it would take four years to get there!

SO MUCH TO LEARN

The definition of a planet changed in 2006. For years, Pluto was believed to be the farthest large object from the Sun. In October of 2019, NASA announced that 20 new moons had been discovered orbiting Saturn.

Because of the phenomenal size of our solar system, we still have so much to learn. The existence of the Oort cloud is yet to be proven. Astronomers believe there could be hundreds of undiscovered dwarf planets, millions of asteroids, and trillions of comets. With recent advances in technology, we've only just scratched the surface.

All this around just one star, our Sun. There are roughly 250 billion stars in our galaxy. And our solar system, along with the hundreds of billions of others, are all rotating around the center of the Milky Way.

With the vastness of our solar system one thing is for sure: There is much more to be discovered, so keep looking up!

GLOSSARY

albedo (al-**be**-do): how reflective an object is

aphelion (a-**feel**-e-yun): the point in an orbit when an object is farthest from the Sun

apogee (**ap**-uh-jee): the point in an orbit when an object—like the Moon—is farthest from Earth

astronomical unit (AU) (as-truh-**nom**-i-kuhl **yoo**-nit): the distance from the Sun to Earth, about 93 million miles

barycenter (**bar**-i-**sen**-ter): in astronomy, the center of mass around which objects revolve

chromosphere (**kroh**-muh-sfeer): the lower layer of the Sun's atmosphere

contact binary system (**kon-**takt **by**-ner-ee **sis**-tuhm): two objects that are touching and orbiting each other as a system

convective zone (kuhn-**vek**-tiv zohn): the layer of the Sun just outside the radiative zone where particles move through convection cells

corona (kuh-**roh**-nuh): the outer layer of the Sun's atmosphere

coronal mass ejection (CME) (kuh-**roh**-nul mas ih-**jek**-shuhn): an event during which the Sun emits a large number of particles from its surface

crescent (**kre**-sent): the shape formed when more than a new moon but less than a half moon is visible

eccentricity (ek-sen-**tri**-sit-ee): how much an orbit differs from a circle

eclipse (ek-**lips**): an event in which one object passes in front of another, blocking its light

ecliptic (ih-**klip**-tik): referring to the plane of the Sun and the orbits of the planets; referring to the path these objects take across our sky

electromagnetic wavelength (ih-lek-troh-mag-**net**-ik **weyv**-lenth): the distance from one crest to the next (or one trough to the next) of an electromagnetic wave

galaxy (**gal**-uhk-see): a system of millions or billions of stars, planets, and dust, held together by gravity

gas giant (gas **jahy**-uhnt): a large planet composed primarily of gas; another name for any of the outer planets

gegenschein (**gey**-guhn-shahyn): a patch formed when dust on the ecliptic, found on the opposite side of the Earth as the Sun, becomes visible at midnight by reflecting sunlight

geocentric (jee-oh-**sen**-trik): a model of the solar system based on the Earth being the center of the solar system and the body around which everything revolves

geologically active (jee-uh-**loj**-i-kuh-lee **ak**-tiv): when a body, such as a planet or moon, has a heat source within it, causing movement inside the body

gibbous (**gib**-uhs): relating to the shape formed when more than a half moon but less than a full moon is visible

habitable zone (**hab**-i-tuh-buhl zohn): the area surrounding a star where life could exist

heliocentric (hee-lee-oh-**sen**-trik): a model of the solar system based on the Sun being the center of the solar system and the body around which everything revolves

hot spot (**hot** spot): a location on Earth's surface where magma has melted through the crust and formed a volcano (for example, Hawaii)

International Astronomical Union (IAU) (in-ter-**nash**-uh-nl as-truh-**nom**-i-kuhl **yoon**-yuhn): the worldwide agency that makes decisions about space discoveries and classifications

interstellar (in-ter-**stel**-er): between stars

ion (**ahy**-on): a charged particle

lunar (**loo**-ner): relating to the Moon

magnetic field (mag-**net**-ik feeld): the area surrounding a magnet on which the magnet has an effect

meteor (**mee**-tee-er): small rock particles that burn up as they enter Earth's atmosphere, producing a moving light seen in the sky

meteor shower (mee-tee-er **shou**-er): an event during which large amounts of meteors appear to radiate from a specific place in the sky

meteorite (mee-tee-uh-rahyt): a small rock from outer space that has passed through Earth's atmosphere and made it to Earth's surface

meteoroid (mee-tee-uh-roid): a small rock in space

nebula (neb-yuh-luh): a cloud of gas and dust in outer space

nuclear fusion (noo-klee-er **fyoo**-zhuhn): the combining of the nuclei of atoms

penumbra (pi-**nuhm**-bruh): a partial shadow created by an eclipse

perigee (per-i-jee): the point in orbit when an object—like the Moon—is closest to Earth

perihelion (per-uh-**hee**-lee-uhn): the point in orbit when an object is closest to the Sun

photosphere (foh-tuh-sfeer): the outermost layer of the Sun, found above the convective zone

plasma (plaz-muh): the fourth state of matter—made of ionized particles and formed at very high temperatures—in which electrons have been stripped from their orbitals

prominence (prom-uh-nuhns): a stream of gas particles above the surface of the Sun that follows magnetic field lines

protoplanet (proh-toh-plan-it): a collection of matter that has the potential to form into a planet through condensation

radiative zone (rey-dee-ey-tiv zohn): the layer of the Sun just outside its core, where heat radiates, or travels outward

Rayleigh scattering (rey-lee **skat**-er-ing): when some wavelengths are scattered more efficiently as light moves through a substance, such as air

retrograde (re-truh-greyd): backward; in this case, referring to backward rotation or orbit

satellite (sat-l-ahyt): anything that orbits a planet; another name for a moon

SETI (se-tee): a collective term for searches for extraterrestrial intelligence that are focused on looking for a planetary body that might harbor life

shepherd moon (shep-erd moon): a moon that helps a ring stay together around a planet

sidereal day (sahy-**deer**-ee-uhl dey): the time it takes a planet to rotate once from the perspective of distant stars

silicate (sil-i-keyt): any substance that contains the compound silicon dioxide as part of its chemical makeup; silicates make up 90 percent of Earth's crust

solar day (soh-ler day): the time it takes a planet to rotate once according to the position of the Sun in the sky when viewed from the surface of the planet

solar flare (soh-ler flair): a sudden flash of light on the Sun caused by the snapping of magnetic field lines that release a large amount of energy

solar radiation (soh-ler rey-dee-**ey**-shuhn): energy given off by the Sun, which comes in the form of ultraviolet light, visible light, and heat

solar system (soh-ler **sis-**tuhm): the entire system of planets, moons, comets, asteroids, meteoroids, centaurs, and other objects and dust, moving together around the Sun

solar wind (soh-ler wind): plasma streaming outward from the Sun

sublimate (suhb-luh-meyt): to change directly from a solid to a gas without becoming a liquid

sunspot (suhn-spot): a spot on the Sun that appears darker because it is cooler

tectonic plate (tek-**ton**-ik pleyt): a piece of the Earth's surface that makes up continental and oceanic surfaces; tectonic plates are shifted around by the movement of magma beneath them

terrestrial planet (tuh-**res**-tree-uhl **plan**-it): any inner planet with characteristics that are relatively similar to Earth's

thermal erosion (thur-muhl ih-**roh-**zhuhn): when heat causes a substance to move; on some moons, heat causes ices to sublimate, which ultimately causes the movement of dust on the moon

tidal force (tahyd-l fohrs): the force that arises when one object's gravity pulls more strongly on the nearest part of another object than on the most distant part

tidal lock (tahyd-l lok): when the same side of an object in space always faces another object because of tidal force; for example, a moon that always faces the same way to its host planet

totality (toh-**tal**-i-tee): the time during an eclipse when one object is completely blocking the light of another; for example, in a solar eclipse, when the Moon is completely in front of the Sun, blocking all of its light

trajectory (truh-**jek**-taree): the path an object in motion will follow in space

trojan (troh-juhn): a small object in space that shares an orbit with a larger object

umbra (uhm-bruh): a full shadow created by an eclipse

volatile (vol-uh-til): a substance that changes from a solid to a gas at room temperature or below

waning (wey-ning): when less of the lit side of a moon is visible in the sky each night

waxing (wak-sing): when more of the lit side of a moon is visible in the sky each night

yellow dwarf (yel-oh dwarf): a star from the main sequence that is about the size of our Sun

zodiacal light (zoh-**dee**-akl lahyt): light that appears to radiate from the zodiacal constellations on the horizon at dusk or dawn; light caused by sunlight reflecting off space dust found on the ecliptic

RESOURCES

Websites

SpacePlace.nasa.gov

This website has information about solar system bodies for younger readers, in a kid-friendly format.

SolarSystem.nasa.gov

For older readers or those of you who want more in-depth information on solar system bodies, this NASA website is fantastic.

VisitNASA.com/nasa-visitor-centers

NASA has more than a dozen flight facility locations you can visit across the country. Check them out here!

YouTube.com/watch?v=21X5lGlDOfg

Check out NASA's official live stream on YouTube for up-to-date information and live streaming of all launches.

Apps

StarTracker

This free mobile sky map and stargazing guide app is for all the stargazers out there. You will need to grant it access to your location. When you open the app, be sure to locate the ecliptic and check for the current locations of all the planets.

NASA

This free app follows all NASA news, includes a link to NASA's YouTube stream, and provides information about solar system bodies.

INDEX

ABOUT THE AUTHOR

Lisa Reichley has enjoyed teaching middle school science for 12 years. She lives in the woods with her husband, Zeke, and their three children in a cottage they built from a kit. Lisa is an avid science fair sponsor who has taken middle school students as far as the national level.

CPSIA information can be obtained
at www.ICGtesting.com
Printed in the USA
JSHW012017070521
14410JS00003B/3